View from the Vineyard

A Practical Guide
to Sustainable Winegrape Growing

View from the Vineyard

A Practical Guide
to Sustainable Winegrape Growing

Clifford P. Ohmart

The Wine Appreciation Guild
360 Swift Avenue
South San Francisco, CA 94080
(650) 866-3020
www.wineappreciation.com

Managing Editor: Bryan Imelli
Copy Editor: Judith A. Chien
Design and Composition: TIPS Technical Publishing, Inc.
Indexing: Brandon Ring

Consulting Editors:
Timothy E. Martinson, Ph.D., Cornell University,
Keith Striegler, Ph.D., University of Missouri,
Mark L. Chien, Penn State Cooperative Extension,
Edward W. Hellman, Ph.D., Texas A&M University

Picture Credits:
Lodi Winegrape Commission: Figures 6.4, 8.2, 9.1, 11.1,
11.3, 12.1, 12.2, 12.6, 12.8, 12.11, 13.10, and 14.2–14.5
Dale Goff: Cover and Figures 1.1 and 15.1

Library of Congress Cataloging-in-Publication Data
Ohmart, C. P. (Clifford P.)
View from the vineyard : a practical guide to sustainable winegrape
growing / Clifford P. Ohmart.
p. cm.
Includes bibliographical references and index.
ISBN 978-1-935879-90-9
1. Viticulture—Handbooks, manuals, etc. 2.
Viticulture—California—Lodi—Handbooks, manuals, etc. 3. Sustainable
agriculture—Handbooks, manuals, etc. 4. Sustainable
agriculture—California—Lod—Handbooks, manuals, etc. 5.
Grapes—Organic farming—Handbooks, manuals, etc. 6. Grapes—Organic
farming—California—Lodi—Handbooks, manuals, etc. I. Title.
SB398.4.O45 2011
634.8'8—dc23
2011033419

*To Robert van den Bosch, whose pioneering work in IPM
inspired my professional career; and to my son Joel,
who is working for the future of California wine.*

Contents

Part I Understanding Sustainable Winegrowing

Foreword

It was an honor when Cliff asked me to write a foreword for his new book. Cliff and I first met at a National Grape and Wine Initiative meeting where we both represented the interests of viticulture education in the U.S. While I reside in the cooperative extension section of the academic ivory tower at Penn State University, Cliff has capably served the wine industry in various private sector positions such as the director of the Lodi Winegrape Commission. I am reluctant to refer to him as a consultant because his vision and work have a much more global reach and impact. Time spent in Australia only served to enhance his appreciation for different approaches to growing wine grapes and how educational resources can serve the wine industry. He was carrying the integrated pest management banner before it became a buzz word among politicians, academics and even growers. His leadership in IPM was the perfect preparation for his advocacy of sustainable viticulture best practices.

Quite frankly, there may not be another person in our industry who is better qualified than Cliff to write a guide to sustainable wine growing. If you are a serious grower, then you already know who Cliff is, either through his columns in *Wines & Vines* or as editor of not one but two seminal guides to sustainable viticulture: the *Code of Sustainable Winegrowing Practices* and the *Lodi Winegrower's Workbook*. These are foundational documents that have been widely imitated in other wine growing regions.

As a planet and a species we are at a crossroads. We now know that the natural resources we depend upon for our very existence is not inexhaustible. It is truly incumbent upon each inhabitant of the planet, regardless of location or vocation, to make a conscious effort to reduce our impact on natural resources for the welfare of the environment and humanity. This is the global necessity and urgency of sustainability.

Because of their connection to the land, farmers carry a unique obligation to its stewardship and preservation. Feeding the world is a noble endeavor, and agriculture, despite its economic and social importance, is often taken for granted in the developed world. Since the advent of the industrial age the use of technology and chemistry has dominated production agriculture. While it has helped to feed a hungry and growing population, it has not always existed in harmony with nature.

I fear that sustainability has been reduced to a cultural buzzword; but there may not be a more important word in our contemporary vocabulary. Our collective future may depend on our ability to internalize and practice it. For winegrape growers, there is no better person than Cliff to help them define, understand, plan and implement sustainable practices in their vineyards.

The wine industry is unusual in agriculture in the number of people who enter it with no farming experience. They tend to be professionals from other careers who have a fascination with wine. They have visited Napa and Bordeaux and heard the terms "sustainable" and "organic" and want to "do it" but they don't really know what to do. This book tells them in clear and specific language what they need to do and why.

As an industry, we are lucky to have someone with the personal experience in sustainability who can take the overwhelming amount of information, data, ideas and recommendations, and then synthesize, organize, and write about them in such a clear and concise fashion. *View from the Vineyard* is an outstanding primer and practical guide to the use of responsible and effective viticultural practices. As educators, Cliff and I know that learning is continuous, especially in an ever changing world. His logic of the continuum of sustainability from "not sustainable" to "completely sustainable" is a novel approach that requires constant learning and implementation as long as there is a vine to cultivate or a breath to take. It highlights the challenge of sustainability to the responsible grower. From a broad overview of Steiner's biodynamic philosophy to specific regulations that govern the use of diesel engines on farms, Cliff covers the realm of sustainability in his personal style.

In extension education impact is everything. We make recommendations that will have a positive impact on the communities and industries we serve. This book will impact commercial vineyards across the U.S. It is a user friendly guide to sustainability in wine growing that should be required reading for every grape grower. Cliff makes it clear that sustainability is a dynamic discipline and I have little doubt that we will be hearing from him again.

—Mark L. Chien
Statewide Viticulture Educator
Penn State Cooperative Extension

Preface

What is sustainable winegrowing? Why should I care about it? How do I practice it in my vineyard? These are all very important questions that many winegrape growers have pondered over the last few years. However, they are not easily answered due to the complex nature of sustainable farming and the debate currently under way as to what is sustainable and what is not. The intent of this book is not to provide the definitive answers to these questions. Rather it is to provide ideas and perspectives to consider as one attempts to answer them.

I like to think of sustainability as a continuum from "not sustainable" on one end to "very sustainable" on the other. However, if the benchmark for sustainability is an environment that is undisturbed by human activity, then no farming approach will ever achieve complete sustainability. The act of farming, no matter how "natural," leaves an environmental footprint. One image that may help you grasp the challenges presented by sustainable farming is to think of sustainable farming as existing in a world where the horizon is always receding. While advances are always being made to help one become more sustainable, new issues and information arise that make the goal of sustainability farther away. This might be a discouraging thought. However, I think a reasonable goal is for each of us to continually improve in sustainability over time. In other words, move along the sustainability continuum. The topics discussed in this book will hopefully provide the reader with tools to accomplish this goal.

The foundation of my understanding of sustainable winegrowing was laid during my undergraduate and graduate training. I received a Bachelor's Degree in forest biology, specializing in entomology, from the State University of New York College of Environmental Science and Forestry. During my four years there I received a sound training in plant and animal ecology, soils, botany, zoology, entomology, plant pathology, geology, limnology, physics, chemistry, and mathematics. As a graduate student at the University of California Berkeley, I specialized in insect population ecology and various aspects of integrated pest management. I developed my ideas on the various topics related to sustainable winegrowing over a career that spanned three different but interrelated paths. The first was as a scientist doing basic research on insect pest management: how insect populations changed over time and what factors affect the dynamics of insect populations, such as natural enemies, plant chemistry and physiology, and their physical environment. The results of this work were published in peer-reviewed scientific journals. The second was as a private consultant working with growers to develop integrated pest management programs for pests of their orchard crops. The results of this work were solutions for real world pest problems to keep my clients satisfied. The third was somewhat in between these two paths, working with a large group of winegrape growers to develop an area-wide sustainable winegrowing program. The results of this work were a grower education program, the *Lodi Winegrower's Workbook* (a self assessment of sustainable winegrowing practices), and the *Lodi Rules for Sustainable Winegrowing*, a third-party-certified sustainable winegrowing program. The common thread through all of these paths is science. While some sustainable winegrowing discussions involve spirituality and/or values, which are important guides in our everyday lives, I am a firm believer that growing quality winegrapes sustainably is based on sound ecological principles and understanding ecosystem dynamics. It is through the understanding of these principals and properly applying them in the vineyard that one will progress along the continuum of sustainable farming.

I have been writing a bimonthly column on the topic of sustainable winegrowing titled "Vineyard Views" for *Wines & Vines* magazine since 1998. This has given me a wonderful opportunity to contemplate many aspects of sustainable winegrowing and formulate my views on the topic. Many of the ideas discussed in this book were developed and matured during the writing of these columns.

Acknowledgments

I could not have written a book on sustainable winegrowing without the help and support of a great many people, to whom I owe a great debt of gratitude. First and foremost, I must thank the Lodi winegrape growers who welcomed me into their community even though I had no experience in vineyards when I arrived in 1995. Through their warmth, generosity, and willingness to share, I learned most of what I know about viticulture. I also must thank Mark Chandler, the Executive Director of the Lodi Winegrape Commission, who supported my ideas and work, and gave me the freedom to develop the programs I felt would be the most beneficial to the Lodi wine community. Stuart Spencer, the Program Manager at the Lodi Winegrape Commission, was also a great resource. I was extremely fortunate to have hired and worked with two of the most talented viticulturists I know, Steve Matthiasson and Chris Storm. I learned a great deal from them during their respective tenures at the Lodi Winegrape Commission. Steve helped put together the first edition of the *Lodi Winegrower's Workbook*, and Chris helped create the *Lodi Rules for Sustainable Winegrowing Program* and helped put together the second edition of the *Lodi Winegrower's Workbook*. I would never have been able to write a book on sustainable winegrowing if Karen Ross, then President of the California Association of Winegrape Growers, had not suggested my name to Phil Hiaring in 1998 as a possible writer for *Wines and Vines*. My more than 12 years of writing bi-monthly columns have given me a wonderful opportunity to explore the various aspects of sustainable winegrowing and try out ideas in a public setting. Over time, while writing my column, I formulated the thoughts that would become this book. I also want to thank Jeff Dlott for the many hours of thoughtful discussion and debate about all matters related to sustainable winegrowing, for collaborating on projects I carried out in Lodi, and for providing opportunities to work on statewide sustainable winegrowing projects in California. Once the book was drafted, Drs. Ed Hellman, Texas A & M University; Mark Chien, Pennsylvania State University; Tim Martinson, Cornell University; and Keith Striegler, University of Missouri patiently and painstakingly reviewed the manuscript providing insightful suggestions for improvement. Finally, I want to thank my wife and life partner Jeri for traveling with me from California to Australia and back, providing support and inspiration in my professional and personal life.

—Cliff Ohmart
Davis, California

List of Figures

Part I

Understanding Sustainable Winegrowing

1

Defining Sustainable Winegrowing

Based on my experience working with growers to develop a regional sustainable winegrowing program, there are three challenges to overcome: 1) Defining sustainable winegrowing, 2) Implementing it in the vineyard, and 3) Measuring its effects on winegrapes and wine, the vineyard, the surrounding environment, and the grower's economic bottom line. Interestingly, I also found this to be the case working with growers to develop integrated pest management programs for their orchard crops, of course replacing the words "sustainable winegrowing" with "integrated pest management." This book is an attempt to provide the reader with information to meet these challenges.

WHAT IS SUSTAINABLE WINEGROWING?

If I were to assemble 50 people in a room, including growers, environmentalists, scientists, and government regulators, and ask them to define sustainable farming, I would likely get 50 different definitions. One of the difficulties in creating a definition is figuring out its boundaries. Sustainability involves all activities undertaken on the farm and how they affect its economic viability, its environmental impacts, and its impact on all aspects of human resources, from employees to the surrounding community. Opinions can differ greatly on which ones are the most important. Another reason is that for some, certain aspects of sustainable farming are value-based, while others, like me, think that sustainable winegrowing should be science-based. And finally, not all farming practices can be evaluated by applying economics, environmental impact, and social impacts in equal measure.

How do I define sustainable winegrowing? I prefer to use the ideas formulated by the developers and practitioners of the sustainable farming and Integrated Pest Management (IPM) movements. It is a systems view of winegrape growing that considers soil building as the foundation, minimizes off-farm inputs, emphasizes

Figure 1-1   Vineyard in Lodi, California with permanent "wall to wall" perennial grass cover crop and oak trees within the vineyard (© Dale Goff).

economic profitability, and concerns itself with environmental health and social equity. The California wine industry, more than any other agriculture sector in the US, has focused on developing an industry-wide sustainable farming program. During the process they defined sustainable winegrowing as "growing and wine-making practices that are sensitive to the environment (environmentally sound), responsible to the needs and interests of society-at-large (socially equitable), and are economically feasible to implement and maintain (economically feasible)" (Dlott et al. 2002). This definition is often referred to as the three "E's" of sustainability and is the one that I use in my presentations to growers and other groups.

When trying to come to grips with a complicated topic like sustainable winegrowing, I find it useful to study how the idea evolved. Sustainable farming, organic farming, Biodynamic® farming, and integrated pest management grew out of the same roots, so to speak. So let's examine how each developed over time.

Evolution of Organic Farming

When defining sustainable farming we need to look at the history of organic farming, since they share a common ancestry. The present paradigm of organic farming began as a melding of several different schools of thought that were supported by European and English scientists active in the 1920s, 30s and 40s. As one would expect, opinions differ as to who really started the organic movement, with at least two people, both British, being bestowed the title of founder: Lady Eve Balfour and Sir Albert Howard. Both practitioners emphasized the role of a healthy, fertile soil in viable agriculture. Howard developed many of his ideas prior to World War II in India, where he was trying to meet the challenge of improving farmers' yields in order to feed a rapidly increasing population. He believed that the best way to increase food productivity at a moderate cost was to return the organic by-products of crop production as well as animal manures to the soil. Howard also had concerns about the changes in soil chemistry caused by the use of synthetic fertilizers and the use of chemical pesticides to solve all pest problems (Francis and Youngberg 1990; Rodale 1973).

The Emergence of Sustainable Agriculture

In the 1950s and 1960s another movement, called the Green Revolution, evolved to meet the challenge of providing food for a rapidly expanding world population. This movement met the challenge from a direction that was diametrically opposed to that of organic farming. It emphasized genetically enhanced plant varieties[1] and high energy off-farm inputs such as mechanization, synthetic fertilizers, and pesticides. In time this movement became "conventional" agriculture and resulted in high food production at a low cost to the public (Parayil 2003). As this movement developed, some people became concerned that this type of agriculture could not be sustained in the long term. They felt that although the cost of food production was low, the dollar value of food produced with conventional agriculture did not reflect the true cost from an ecosystem and societal perspective. The true cost takes into consideration issues like air pollution from producing and using fossil fuels, soil degradation due to intense cultivation and use of synthetic fertilizers, habitat destruction, air and ground water contamination with fertilizers and pesticides, and the steady decrease of the farmer population as small family farms were out-competed

---

1. The breeding referred to here we would now define as conventional breeding.

by large corporate farms. These concerns over the long-term viability of conventional agriculture accelerated the evolution of the sustainable agriculture movement, which owes many of its farming approaches to the organic farming movement.

Organic farming originally developed as a farming paradigm, but over time some realized it could be a way to add value to produce and other agricultural products through marketing. In other words a price premium could be achieved for organically grown food and other farm products. This necessitated the development of certification programs to codify organic farming practices and verify they were being followed. Unfortunately, as is the case with some marketing programs, a number of farmers bent or broke the rules of organic farming or even came up with their own interpretation of what constituted organic farming in order to get a price premium for their produce. Other terms like "natural farming" started showing up on produce labels and at farmers' markets. A point was reached in the US when some members in the supply chain, particularly the retailers, felt a federally recognized organic farming program was necessary to reduce the confusion around food labeling and to ensure the credibility of the organic label. They approached the US Department of Agriculture about the problem and the end result was the development of the National Organic Program (NOSB 2007).

With the development of organic certification programs, organic farming became codified and easily distinguished from other farming strategies. However, as yet no one has attempted to codify sustainable agriculture at a national level, although recently the American National Standards Institute (ANSI) has convened a committee to develop sustainable farming standards for all of US agriculture. Due to the absence of a nationally recognized code of practices for sustainable agriculture there is an active debate among academics, farmers, environmentalists, and others as to what defines sustainable agriculture and what practices are considered sustainable. Some consider it to be a philosophy, others consider it to be a guideline for determining farm practices, some view it as a management strategy, and others argue about whether it is strictly related to farm production or also encompasses sociological issues.

In 1989 the American Agronomy Society adopted the following definition for sustainable agriculture: "A sustainable agriculture is one that, over the long term, enhances environmental quality and the resource base on which agriculture depends; provides for basic human food and fiber needs; is economically viable;

Figure 1-2   Cabernet Sauvignon vineyard near St. Helena, California in March, with a mustard cover crop.

and enhances the quality of life for farmers and society as a whole." The Sustainable Agriculture Research and Education Program at the University of California, Davis (UC SAREP) emphasizes that sustainable agriculture integrates three main goals—environmental health, economic profitability, and social and economic equity. UC SAREP also points out that a systems perspective is essential to understanding sustainable agriculture. Farming does not operate in a vacuum. Each farmer's field is part of a complex community ecosystem, which in turn can affect or be impacted by global economics and even global ecological processes (e.g., ocean temperature cycles such as El *Niño*). A systems perspective involves viewing multiple factors when considering field and farm-level decisions.

In 1987 the World Commission on Environmental Development published a definition of sustainable development that is another definition some use when discussing sustainable farming. The report stated that sustainable development is "...development that meets the needs of the present without compromising the ability of future generations to meet their own needs."

WHERE DOES INTEGRATED PEST MANAGEMENT FIT IN?

Integrated Pest Management (IPM) was developed in the late 1950s to deal with some of the pest problems that in many ways can be attributed to the farming practices developed during the Green Revolution (Stern et al. 1959). The use of genetically enhanced plant varieties and over-reliance on pesticides to solve pest problems resulted in pesticide resistance and secondary pest outbreaks, as well as environmental contamination. It is important to note, however, that IPM was initially developed more from a problem-solving, economic imperative rather than from a need to reduce off-farm inputs and to protect the environment. The formalization of IPM occurred several years before the publication of Rachael Carson's *Silent Spring*, a seminal book that documented the detrimental effects of pesticides on the environment, particularly their impacts on birds.

IPM in the US came about because there were several crops, particularly alfalfa and cotton, which had developed unmanageable pest problems due to pesticide resistance and insecticide-induced secondary pest outbreaks. Scientists working in these crops realized that the over-use of pesticides had brought them to this point and that the only way out was to integrate several control strategies and reduce reliance on pesticides. It turns out that IPM strategies fit right in to the paradigm of sustainable agriculture and the environmental movement and thus has become an integral component of both (see Chapter 13 for further discussion).

Like sustainable farming, IPM has not been codified, and therefore can mean many things to many people. A multitude of definitions has been proposed. I personally prefer the following definition: "IPM is a sustainable approach to managing pests by combining biological, cultural, and chemical tools in a way that minimizes economic, environmental, and health risks." I like to think of IPM as a problem-solving tool. It is an approach to managing pest problems, just as sustainable agriculture is an approach to farming. Like sustainable farming, a helpful way to understand it is to visualize it as a continuum from no IPM on one end to a high level IPM on the other.

Although the concepts of sustainable farming, organic farming, and IPM have been around for a long time, they are still often misunderstood or misinterpreted according to one's bias. For example, a farmer dedicated to organic farming may not have the same definition of sustainable farming or IPM as someone who does not restrict their farming to organic methods. In any case, it is important to realize that pest problems may still arise, even when practicing sustainable or organic farming and/or using IPM for managing pests problems. That is because most crops are exotic (i.e., non-native) to the farms on which they are grown, and most pests on these crops are non-native, too. Moreover, many of the plant parts we harvest for food contain the highest concentration of nutrients and carbohydrates, which make them not only very useful for us but extremely attractive to many other organisms. This creates a potentially unstable ecological situation regardless of the type of farming being practiced. There are some crop/pest systems that are inherently unstable and crop damage is unavoidable without some outside intervention. A good example is codling moth in many orchard crops. Despite years of research in introducing natural enemies, developing mating disruption programs, and other sustainable techniques, it is still one of the major pests of many orchard crops. Pests can even get out of hand in some fairly undisturbed, "natural" ecosystems, as illustrated by periodic destructive epidemics of forest insects in certain forest ecosystems (e.g., Elliott et al. 1998). (See Chapter 13 for further discussion on IPM.)

THE SUSTAINABLE WINEGROWING CONTINUUM

I think visualizing sustainable winegrowing (and sustainable farming) as a continuum, from less sustainable on the one hand to more sustainable on the other, is very helpful when trying to understand the sustainability paradigm (Figure 1-3). If an undisturbed natural system is the benchmark for complete sustainability, one must realize that no farmer will be completely sustainable, because the act of farming disturbs the natural system no matter how sustainable are the practices (see the next chapter for a detailed discussion of this idea). Therefore, the goal of sustainable winegrowing should be continual improvement, in other words moving along the continuum toward a higher level of sustainability.

In a real sense, sustainable farming is like being on a journey one will never finish, because complete sustainability is not possible. This thought can be very discouraging, since as humans we want to reach an end point when we are doing

Less Sustainable   More Sustainable

Figure 1-3   Sustainable winegrowing viewed as a continuum.

something or where we can classify our actions as a binary "yes" or "no." I encountered a statement recently while listening to a National Public Radio interview with Van Cliburn, where he said the world of art is one where the horizon is always receding. I thought this statement was also a perfect way to express of the world of sustainable farming. In other words, one will never reach complete sustainability because there is always room for improvement. Moreover, new issues continue to arise, such as climate change due to increases in atmospheric carbon dioxide, which show us that our sustainability horizon is farther away than we thought.

It is sometimes difficult to use the 3 E's (economically feasible, environmentally sound, and socially equitable) yardstick when evaluating the sustainability of an individual farming practice. For example, it is difficult to talk about the social ramifications of releasing a parasite to control vine mealybug or the environmental soundness of doing a team building exercise with your employees. However, the sustainability of a farm is measured by examining the sum total of all the practices implemented on the farm using the 3 E's.

And finally, economics is going to dictate what sustainable practices can be implemented. For example, the practices being implemented in a vineyard where the grapes are being sold at $400 per ton are going to be quite different from those that can be implemented in a vineyard where the grapes are being sold for $4,000 per ton.

REFERENCES

American Society of Agronomy. 1989. Decisions reached on sustainable agriculture. Agron News January. p 15.

Dlott, J., C. P. Ohmart, J. Garn, K. Birdseye, and K. Ross, eds. 2002. *The Code of Sustainable Winegrowing Practices Workbook*. Wine Institute & Calif. Assoc. Winegrape Growers. 477pp.

Francis, C. A., G. Youngberg. 1990. *Sustainable Agriculture—An Overview. In:* Sustainable Agriculture in temperate zones. Francis, C. A., C. B. Flora, and L. D. King, eds. John Wiley & Sons, N.Y. pp. 1–12.

National Organic Standards Board (NOSB). 2007. *Policy and Procedures Manual*. Nat. Org. Stand. Board, Washington, DC. 67pp.

Parayil, G. 2003. "Mapping Technological Trajectories of the Green Revolution and the Gene Revolution from Modernization to Globalization." *Research Policy* 32(6):971–990.

Rodale, R. 1973. "The Basics of Organic Farming." *Crops and Soils* 26(3):5–7,30.

Stern, V. M., R. van den Bosch, and K. S. Hagen. 1959. "The Integrated Control Concept." *Hilgardia* 29:81–101.

UC-SAREP. 2008. University of California Sustainable Agriculture Research and Education Program Website: http://www.sarep.ucdavis.edu

World Commission on Environment and Development. 1987. *Our Common Future*. Oxford University Press. 398pp.

2

Sustainable Winegrowing Viewed from Different Perspectives

HOW "NATURAL" IS FARMING?

I often ask that question in lectures for at least two reasons. The first is that it gets people's attention. The second is that it usually gets people thinking about farming from an ecological perspective. The downside of asking it is that many farmers get really annoyed because they see farming as the ultimate natural process. If one sees people as a part of the ecosystem, which I do, then farming is as natural as anything else people do. However, some consider people as separate from other animals and therefore either separate from the ecosystem or at least a special case. Where farming fits all depends on one's perspective.

Humans began in this world as hunter-gatherers, living off the land just like all the other animals. They were omnivores, meaning they ate pretty much anything they could get their hands on and digest properly, both plant and animal tissue. Since it was the first form of survival, an argument could be made that it was the most natural one. Although in some situations hunter-gatherers had a very big impact on the environment. For example, the aboriginal tribes in Australia regularly burned large areas of the continent to create an environment more conducive to producing the food they desired.

The success of the hunter-gatherer lifestyle was completely dependent on the vagaries of the local climate, which in turn affected the local plant and animal populations. In some years food was plentiful but in others it was not. At some point humans began to cultivate crops because it was a more reliable way to ensure a regular supply of food. As soon as they gave up the hunter-gatherer lifestyle to stay put and grow their own food, producing more than they could consume in one sitting, they left what might be considered a more natural way of life and headed down the less-than-natural road of farming.

From this perspective farming can be viewed as an unnatural act for several reasons. For one thing, virtually all of the crops we grow, whether organically or conventionally, are varieties that are exotic. That is, they are not native to (did not evolve in) the area in which they are now grown. In some cases this has had dire consequences. Take winegrapes, for example. Virtually all of the varieties that are grown in the US evolved outside North America, primarily in Europe. It turns out that grape phylloxera, an aphid that evolved in North America on native grapes, devastates European grape varieties because they have no natural immunity, not having evolved in the presence of this insect. The only way to successfully grow European grapes in areas where phylloxera occurs is to grow them on North American grape rootstock. There are many other cases like this in agriculture throughout the world, where a crop was imported and it proved to be susceptible to a local insect or disease.

Not only are we growing crops where they are not native, but we have also altered plants as result of centuries of genetic selection, so that they provide the maximum amount of the plant part we desire, whether it is the root, flower, fruit, or leaf. This, too, is unnatural. For example, fruit and roots that we grow and consume are packed with many more nutrients than their ancestors had. This is true even for heirloom plant varieties. The consequences of this genetic selection have had a dramatic effect not only on our evolution but also on certain species that feed on these plants. Not only do these plants provide us with more nutrition in the parts we eat, they also are more nutritious to the insects and diseases that feed on these plant parts. We have improved the plants for our benefit but also, ironically, for the benefit of these other species which we call pests. In many cases we have created our own pest problems.

Another way farming can be viewed as an unnatural act is by realizing that regardless of the size of a farm, whether it is a 10,000-acre corporate giant or a one-acre organic patch, natural habitat was destroyed when it was established. Land had to be cultivated, whether by hand or by machine, and pre-existing plants had to be removed. Furthermore, no matter how diversified is the resultant farm, it is still more ecologically simplistic (less bio-diverse) than the natural habitat it replaced.

Because farming alters an ecosystem from its natural state, it has significant environmental impacts. Some perceive the significance of these problems and their causes differently from others and therefore develop different ways of reacting to them. Two human characteristics that can get us into conflict are denial and the need to feel superior to others. I often encounter them when reading about or hearing people discuss the impacts farming has had on the environment and society and the two major farming paradigms. I am referring to the organic vs. conventional farming debates. Some organic farmers and their supporters view it as the most sustainable way to farm, while some conventional growers say things like "my family has been farming for several generations; now that is what I call sustainable!" In both cases they are not admitting the fact that farming has a significant impact on the environment no matter how it is done.

So where does this leave us? I think we all need to admit that no matter how naturally one farms, it is still a process that highly modifies the environment in which it occurs. In other words, it leaves an environmental footprint. Nevertheless, it is subject to ecological processes and should be thought of and evaluated in ecological terms. Farming is not really natural or unnatural. It is something that we have to do to produce food and we should endeavor to practice it using strategies that are as sustainable as possible.

How Are Organic and Sustainable Farming Related?

Food and fiber grown using organic practices have entered main-stream consumer markets. For example, organically-grown produce and meats can be found in most grocery stores and make up a large percentage of produce at farmers' markets. I even saw an organic mattress advertised recently in the local newspaper. As a result, most people have heard the word "organic" and have a perception of what it means. On the other hand, marketing the idea of producing food and fiber using sustainable practices is a much newer phenomenon. Many people are not familiar with the term "sustainable" and are not likely to have much of a perception of its meaning. Usually when I talk with people about sustainable winegrowing their first question is "How does it differ from organic?"

Several wine grower groups in the US have very active sustainable winegrowing programs that encourage their members to use sustainable practices in their vineyards. Furthermore, a few are marketing their winegrapes as grown using sustainable practices, for example, the Lodi Winegrape Commission through their *Lodi Rules for Sustainable Winegrowing* (www.lodirules.com), Fish Friendly Farming (www.nswg.org/n3.fishfriendly.htm), Oregon LIVE (*www.liveinc.org*), Salmon Safe

Figure 2-1   Logo for the Lodi Rules for Sustainable Winegrowing Certification Program.

(*www.salmonsafe.org/wine/index.cfm*), and Central Coast Vineyard Team Sustainability in Practice program (www.vineyardteam.org). To be successful in the marketplace, wines labeled as made from grapes grown using sustainable farming methods must be differentiated from other wines, including those made from organically grown grapes. Since most consumers have a perception of what "organic" means, these new programs must create a consumer perception of what "sustainable" means to be successful. Making consumers aware of what sustainable means is a huge challenge and it starts by understanding how organic and sustainable are related and how they differ.

THEORY VS. PRACTICE

Farming paradigms, such as organic or sustainable, can be divided into two basic components, the theory or principles that underpin the foundation of the paradigm and the farming methods that are used to put the principles into practice.

Theory

When one examines the principles of sustainable vs. organic farming they look very similar. That is because they both trace their history to the same roots, as discussed in Chapter 1. The paradigm of sustainable farming has a much longer history than that of organic farming, but its birth date is very indefinite. One historical synopsis of sustainable agriculture by R. MacRae *(www.eap.mcgill.ca/AASA_1.htm)* concludes that it has been a part of English farming for several hundred years and evolved from three perspectives: as a system of production to achieve food self-reliance; as a concept of stewardship; and as a vehicle for sustaining rural communities. Organic farming, as we understand it today, has a much more definite and recent beginning. In a review of the history of organic farming, M. H. R. Viandes *(www.mhr-viandes.com/en/docu/docu/d0000153.htm)* points to Rudolf Steiner's Biodynamic farming concept, proposed in a series of lectures in 1924, as one of the beginning points of organic farming, and it will be discussed in greater detail in Chapter 3. As already mentioned in Chapter 1, credit was also given to Sir Albert Howard in England, where during the 1940s he promoted his theory of the importance of humus and soil fertility as the foundation of good agricultural practice. It is important to note that at this time the word "organic" was not yet associated with what we now think of as organic farming. MacRae reports that the first widespread use of the word "organic" as it relates to farming was in a book Look to the Land by Lord Northbourne published in 1940. He used the word in a broad sense to describe farming systems that focused on the farm as a dynamic, living, balanced, organic whole, or an organism. The first time the word organic was used widely in the US, according to MacRae, was in the 1950s by J. I. Rodale, founder of Rodale Press.

When one looks at the various definitions of organic farming, they all discuss the same broadly focused principles, such as producing sufficient quantities of high-quality food and fiber, maintaining and increasing the long-term fertility and biological activity in the soil, maintaining biodiversity on the farm, and using renewable resources in the process of farming. For example, here is the definition adopted by the National Organic Standards Board in 1998: organic farming is "an ecological production management system that promotes and enhances biodiversity, biological cycles, and soil biological activity. It is based on minimal use of off-farm inputs and on management practices that restore, maintain and enhance ecological harmony."

Definitions of sustainable agriculture contain the same principles, and these were discussed in Chapter 1.

Practices

Farming is achieved through making many decisions that result in implementing a set of practices, in other words putting theory (or definitions) into practice. While

Figure 2-2   Crimson clover cover crop that adds nitrogen to the soil and provides nectar and pollen to insect predators and parasites.

some of the practices used in organic and sustainable farming are similar, as sustainable agriculture becomes more codified the differences in the systems may become more apparent.

First, when comparing the two systems, it is important to note that one major aspect setting organic farming apart from sustainable farming in the US is the existence of a nationally recognized set of farming practices that one must adhere to in order to label food or fiber as being "organic." These practices are certified by the US Department of Agriculture. An equivalent national system does not exist for sustainable farming. It is therefore tricky to compare practices of the two systems because the organic system is codified while the sustainable system is not. Without agreed upon standards for sustainable farming, which practices are "officially" sustainable is open to debate. Clearly, one of the advantages of national organic standards is that while farming groups around the country may disagree somewhat on the wording that describes the organic paradigm, everyone must adhere to the same set of farming practices to achieve organic certification under the National Organic Program.

The establishment of national organic standards is relatively recent (October 2002). As organic farming matured, the need for a nationally recognized set of standards became essential, because people were labeling food and fiber products as organic but in some cases some of the practices used differed from one farmer to another. Moreover, some less than scrupulous producers were using some non-organic practices and still labeling their produce as "organic." Eventually the industry realized that although the development of national standards would increase the regulatory burden on organic farmers, it was necessary to do so to protect the integrity of the organic label.

Despite the absence of a nationally agreed upon set of sustainable farming practices, enough has been written about sustainable practices and the existence of local or regional certification programs provide enough of a pattern to make some useful comparisons between the two systems.

Practices that organic and sustainable farming have in common are ones that minimize soil erosion, do not contribute to contamination of surface and ground water, and build soil fertility using cover crops and adding organic soil amendments such as manures or compost.

Probably the most notable difference between organic certification and the existing sustainable farming certification programs is that organic farmers cannot use synthetic pesticides or fertilizers, or any other synthetic inputs, with a few exceptions. While most sustainable certification programs have a pesticide "do not use" list as well as a list of other pesticides that can be used but with restrictions, most allow the use of some synthetic pesticides as well as fertilizers.

In some circles, the use of any synthetic materials is not considered sustainable. However, this is a very complicated issue and I will not attempt to resolve it here. I will say that in evaluating whether the use of any chemical, either synthetic or organic, on the farm is sustainable, one needs to consider the energy it took to produce it, the energy it takes to apply it, and its environmental toxicity, including worker exposure effects. The fact that a material is naturally derived is not sufficient to declare it usable in a sustainable farming system. There are some organically approved pesticides that have greater toxicity to non-target organisms than many of the newer synthetic pesticides.

It is clear that pesticide use is a very important focus of both farming systems but is being handled differently by each one. As has been mentioned already, the only pesticides that can be used on a certified organic crop are ones that are naturally derived, with a few exceptions, and that are approved for use by the National Organic Standards Board. On the other hand, one sustainable farming certification program, Healthy Grown™ (*www.wisconsinpotatoes.com/fresh_healthy.htm*), developed by a collaboration of the Wisconsin Vegetable and Potato Growers Association, the University of Wisconsin, and World Wildlife Fund, has developed an

Figure 2-3   Logo for the Healthy Grown sustainable farming certification program for Wisconsin fresh market potatoes.

innovative way of measuring pesticide impact based on the type and quantity of each pesticide used. To be certified a field cannot exceed a threshold of pesticide impact units in a given year. An impact coefficient is calculated for each pesticide based on mammalian toxicity, aquatic toxicity, and toxicity to pest natural enemies. The impact units are then calculated by multiplying the impact coefficient for each pesticide by the amount applied to each field acre. Experience has demonstrated a chemical may be safe per se but if too much is used it can have adverse impacts anyway. This measurement system is an attempt to take this into consideration and uses quantitative measurements to assess the impacts of all types of pesticides, whether synthetic or organically approved, using the same scale (Benbrook et al. 2002).

Protected Harvest, a non-profit third-party certification organization, grew out of the Healthy Grown program and has gone on to accredit and certify other sustainable farming programs growing other crops, requiring that a similar pesticide impact model be developed for each one. One example is the *Lodi Rules for Sustainable Winegrowing* (Ohmart et al. 2006, *www.lodirules.com*).

At the time of the writing of this book a next generation pesticide risk model is being developed by a team of scientists funded by a grant from the Natural Resources Conservation Service. The name of the model is Pesticide Risk Mitigation Engine or PRiME. It will measure a pesticide's risk based on 15 different indices, including avian reproduction, avian toxicity, earthworms, aquatic algae, small mammals, bees, daphnia, human dermal, human dietary, and VOC potential (www.ipminstitute.org/prime).

Another important issue that differentiates some sustainable certification programs from organic certification is ecosystem management. National organic certification does not have any requirements relating to wildlife habitat and other

Figure 2-4   Logo for Protected Harvest, the non-profit organization that certifies sustainably grown food.

ecosystem management issues, while some winegrape certification programs, such as Fish Friendly Farming, Oregon LIVE and the *Lodi Rules* have these require-ments. Furthermore, the *Lodi Rules* program also includes human resource stan-dards, another issue not covered by organic certification.

As previously indicated, sustainable farming programs are embracing the most recent environmental concerns such as habitat destruction, water use efficiency, energy consumption, labor issues, and production of greenhouse gases. Even though sustainable farming has a long history, it has a short history when it comes to certification. The advantage of this is that when a group creates a sustainable cer-tification program they can incorporate farming standards that deal with these issues. Organic farming has been codified for a relatively long time, and when the codification was developed the environmental concerns described above were not as apparent as they are now. The result is that organic farming standards do not deal with many of these issues. The organic community is now discussing how to meet this challenge.

3

A Brief Synopsis of Rudolf Steiner and Biodynamic Farming

Many of us have a basic understanding of what organic farming is, and quite a few winegrape growers in California adhere to its standards. The term sustainable farming is also familiar to many of us. On the other hand, most of us have little or no knowledge of Biodynamic farming. Some growers have never heard the term before, while others, when hearing the word, envision voodoo dolls, chanting, applying strange potions to the soil, and occult practices. I have a feeling that we are going to hear more about this type of farming, and some practitioners claim that farming Biodynamically "can cure any insect, disease, yield or quality problem on any crop, anywhere." (*Wine Business Monthly* August 2000 pp. 45-54). With rhetoric as absolute as this it is important to become familiar with Biodynamic farming so that it can be discussed in a coherent fashion. It is not possible in a few pages to thoroughly cover Biodynamic farming. However, I can give a brief history of the origins and founding principles of biodynamic farming so as to put it into perspective with other farming strategies.

To understand Biodynamic farming one must understand a bit about Rudolf Steiner. Many people are responsible for the evolution of most farming paradigms such as organic and sustainable farming. However, the principles and practice of Biodynamic farming are attributable to only one person, Rudolf Steiner. Biodynamics can be traced directly back to a series of eight lectures developed and presented by him in June of 1924 to a group of European farmers who came to him for advice on soil fertility problems, degenerate seed strains, and the spread of animal disease. Steiner died in early 1925 and others have carried on his work. To fully appreciate what is behind Biodynamic farming, it is important to understand Steiner the scientist.

Figure 3-1   Rudolf Steiner.

DR. RUDOLF STEINER, "SPIRITUAL" SCIENTIST AND RESEARCHER

Rudolf Steiner was born in 1861 in a small town in what is now Croatia. He went to technical school as a youth and was well grounded in the natural sciences. Out of his own interests he began reading a great number of philosophy books. He became convinced that it was only through the philosophical method that the material and spiritual worlds would be bridged. Throughout his advanced studies in math, natural history, and chemistry he continued his keen interest in the work of contemporary philosophers. He saw a constant interplay between the material and spiritual worlds. He obtained a Ph.D. in 1891 and taught history, German literature, and the history of science in Berlin for several years. In 1902 he declared in a lecture that his life's aim was to found new methods of spiritual research based on science.

In a biography of Steiner, Gilbert Childs (1995) writes "Steiner was an explorer of worlds closed to the ordinary powers of sense-perception, and few were capable of following him." Childs develops Steiner's idea that for spiritual perception we need to develop supersensory organs and claimed that Steiner had achieved such a perceptive ability. Nevertheless, since Steiner was trained as a scientist and dedicated to the investigative standards of scientific research, Childs reported that he strove constantly to apply corresponding rigor to his own investigations. Steiner referred to himself as a "spiritual researcher" and felt that the body of knowledge he accumulated was genuine "spiritual science." He coined the term "anthroposophy" as the name of this science. Steiner defined anthroposophy as "a path of knowledge that strives to lead the spiritual in man to the spiritual in the universe" (Koepf 1976).

Steiner's views were considered by many of his contemporaries to be controversial, and there was strong opposition to them, to the point of threats being made on his life. Some felt he was associated with the occult. He began lecturing on diverse topics such as religion, education, social issues, history and human nature. Many sympathizers began to desert him. However, by January 1905, his adherents considered the depth of his knowledge of the material and immaterial worlds was such that invitations to give lectures poured in and his life work had begun.

Around 1917, Steiner began another phase of his career, devoting his time to putting his spiritual-scientific principles and knowledge to practical use. For example, he was approached by the managing director of the Waldorf-Astoria cigarette factory in Stuttgart, Germany, to direct a school for children of factory employees. To accomplish this he started the Waldorf/Steiner school in 1919 and developed an educational system based on anthroposophy. There are now Waldorf schools all over the world. In 1920 he was asked by a doctor to develop a series of lectures for doctors and medical students on various aspects of human anatomy, physiology, and pathology as well as diagnoses and appropriate remedies, including developing some pharmaceuticals. Then in 1924, one year before his death, Steiner gave his series of eight lectures that became the basis for Biodynamic farming.

THE FOUNDATIONS OF BIODYNAMIC FARMING

Steiner took a holistic approach to farming. He felt that since plants germinate, grow, and produce fruit and are dependent on the sun, earth, air, and water to do so, then literally the whole universe is involved in these processes. Another way to put it is that the yield and quality of crops come about under the influence of two groups of environmental factors: earthly and cosmic. He saw each farm as an individual organism which should be as self-sufficient as possible. For example, a Biodynamic farm should have a diversity of crops and a certain amount of livestock. Because a farm is a living organism, he reasoned that only life-endowed substances should be applied to it. "Dead" materials such as chemical fertilizers should not be used. By this same argument, synthetic pesticides should not be used either. Therefore only organically derived materials should be used in farming, and it is in this aspect that Biodynamic farming has a commonality with organic farming. It is interesting to note that Steiner developed these ideas before synthetic, carbon-based pesticides were invented and widely used.

When someone asked for his views about plant diseases Steiner responded by saying that plants could never be diseased in a primary sense, "since they are the

Figure 3-2   Livestock is a requirement for Biodynamic farming, such as this Highland Cow.

products of a healthy etheric world." He believed they are diseased as a result of diseased conditions in their environment, especially the soil (Koepf 1976).

One important practice that sets Biodynamic farming apart from other farming practices, particularly from organic farming, is the use of nine specific preparations of materials developed by Steiner to add to composts, to the soil, or sprayed on plants, depending on the preparation. The amount of the preparation applied is small because he felt that they worked "dynamically," regulating and stimulating processes of growth. Putting it in present day terms, their primary purpose is to stimulate the processes of nutrient and energy cycling. Steiner gave each preparation a number from 500 to 508 and they are divided into two groups. The first group consists of Nos. 500 and 501 and each is applied in spray form. No. 500 consists of dairy cow manure collected in early autumn, packed into a cow's horn, buried in a pit in biologically active soil for the winter and dug up in the spring. No. 501 consists of ground quartz mixed with rain water to make a paste which is then packed into a cow's horn, ideally from a cow that has calved a number of times but not more than eight years old. The horn is then buried in the late spring in a sunny spot and dug up in late autumn. Both 500 and 501 are made into a spray by mixing the end materials with rainwater. No. 500 is sprayed onto the soil while 501 is sprayed onto plants (Sattler 1992).

Preparations 502 to 508 are made from the following plant substances, respectively: yarrow blossoms, chamomile blossoms, stinging nettle, oak bark, dandelion

Figure 3-3   Cow horn similar to one that might be used for a Biodynamic preparation.

flowers, valerian flowers, and horsetail. Each preparation is made in a very specific way. For example, No. 502 is made from yarrow flowers that are put in the bladder of a red deer stag, suspended in the sun throughout the summer and buried in the ground during the winter. It is then added to a compost pile, along with some of the other preparations, to aid the composting process, resulting in Biodynamic compost. Certain animal parts are used in the other preparations, such as bovine mesentery, bovine intestines, and domestic animal skulls. For more detailed descriptions of Steiner's preparations and their uses see Sattler 1992. Sattler emphasized in his book that little or no result can be expected if a preparation is used on its own. It needs to be used in concert with all of the other Biodynamic principles, processes, and preparations.

Rhythms are also an integral part of Biodynamic farming. It is felt that biological rhythms are connected in some way to cosmic rhythms. For example, Steiner felt that sun spot activity, moon rhythms, and the zodiac all have significant effects on the growth and health of plants. Space does not allow a detailed explanation here, but see Sattler's book for more details.

There has been little scientific research into the efficacy of Biodynamic farming practices. The few studies that have been done have focused on the effectiveness of Steiner's preparations. The results have been mixed, but in some studies it was shown that Biodynamic farming systems have better soil quality, such as higher organic matter, greater microbial activity, more earthworm channels, and lower crop yield when compared to conventional farming systems (Reganold 1995).

Because I was trained in the scientific method and my understanding of how biological systems work is based on ecological theories developed using this method, I have a hard time coming to grips with many of Steiner's ideas and recommendations. Not only did he develop his very unorthodox methods with little scientific justification, they cannot be tested in the normal sense because there is an inherent contradiction in his philosophy. Steiner developed his "spiritual" science as an alternative to the traditional scientific method, so traditional science cannot be used to test the efficacy of Biodynamic farming.

Nevertheless, in reading about Biodynamic farming and in talking with people who practice it, it appears its goals as they apply to winegrape growing are very similar to those of sustainable winegrowing. I find it a useless exercise to infer that one farming paradigm produces the best wines. My experience is that there are award-winning wines produced using each of the four farming paradigms, organic, Biodynamic, conventional and sustainable, and there are bad wines produced using each of them, too. I think the key to producing quality wine is attention to the details of how the vines are growing and achieving the ideal vine balance for a particular vineyard and site. Any farming paradigm that focuses on attention to these details is therefore likely to produce fine wines that reflect the character of where they are grown. However, to say one is better misses the point that the goal in farming is to continually improve in whatever system you are practicing.

REFERENCES

Childs, G. 1995. *Rudolf Steiner: His life and work.* Anthroposophy Press, N.Y.

Koepf, H. H. 1976. *Bio-dynamic agriculture: An introduction.* Anthroposophic Press. NY. 429pp.

Reganold, J. P. 1995. Soil quality and profitability of biodynamic and conventional farming systems: A review. *Amer. J. Alternative Agr.* 10(1):36–45.

Sattler, F. 1992. *Biodynamic farming practice.* Bio-Dynamic Agricultural Assoc. 333pp.

4

Constraints to Implementing Sustainable Winegrowing: The Commodification of the Winegrape

Because sustainable winegrowing is a continuum there is always something more one can do to progress along it. It is important to note that the amount of income a grower achieves per acre constrains how many sustainable practices one can implement. A grower getting $4,000 a ton for their Cabernet Sauvignon can afford to implement a lot more practices than a grower getting only $400 per ton, although as we will see later, implementing some sustainable practices lowers the cost of winegrape production.

Consolidation of wineries world-wide has increased significantly in the last few years, increasing the downward pressure on winegrape prices and threatening to turn winegrapes into a commodity.

Figure 4-1   Corn is one of the most important agricultural commodities.

A commodity is defined by Wikipedia as "a largely homogenous product traded solely based on price." Growing a commodity crop seems like a brutal business since the grower has no control over price yet price dictates whether or not a grower will stay in business. When prices are up the grower makes money, when prices are down the grower loses money. About the only thing over which a grower has control is trying to lower the cost of production per unit of product. Since a commodity is homogenous in quality, a processor can buy it from anywhere and still sell it for the same price. If they can buy it from anywhere they are, of course, going to try their best to get it from the cheapest source. As a result, the pressure on the grower is to figure out how to cut the cost of production per unit of crop so they can become the cheapest source. Unfortunately, in the global marketplace there is always someone somewhere else who can grow it cheaper than you can. So, as a colleague of mine has said many times, commodification becomes a race to the bottom. In other words, someone does it cheaper, you figure out how to do it cheaper, then they do, and pretty soon you are out of business.

I am not suggesting that winegrapes will become a commodity in the true sense of the word, like corn, wheat, soybeans, or cotton. However, I am sure most of you growing winegrapes in the major winegrape-growing regions of the US are aware of how the wine industry is changing and seems to be heading in that direction. The following are some of the signs. Periodically very large crops are produced, such as in California in 2005, which fill all the winery tanks and lower grape prices to growers. Wine consumption in the US continues to grow, making us a target for wine exporting countries from around the world. There has been a significant consolidation in wineries world wide, reducing the number of winegrape buyers and increasing the number of large wineries with the ability make or purchase and move bulk wine around the world. In the US wineries are allowed to blend up to 25% of foreign wine with US wine and still call it "American." With the recent passage of vintage dating, non AVA wines can contain up to 15% of the wine from a different vintage from what is on the label. This may not sound like much but the following numbers put it into perspective. Say you are a large winery and have made wine from 85,000 tons of grapes. With the new law you can blend that wine with 15,000 tons of grapes from another vintage. That blending represents 3,000 to 7,000 acres of grapes, depending on the yield, that are now not needed if the winery has wine left over from a previous year.

Unfortunately, many of California's specialty crops, such as citrus, apples, and cut flowers, have also been severely affected by the globalization of agriculture. With these crops the problem is not related as much to consolidation of processors as it is to the fact that it is cheaper to grow them in other countries, ship them to the

US, and still sell them for less than the same crops produced in California. In these cases the main difference seems to be land and labor costs and possibly low cost inputs. Some would like to think that many of these overseas farming operations may be cheaper but not very sophisticated. However, in talking to people who have visited some of these operations it turns out that they are using all of the latest knowledge and technology.

Is this scenario really sustainable? I would say "No" because I see it as a mining operation rather than something that is renewable. They "mine" land and human resources until the cheap supply runs out, then they move on to other countries where these resources are still cheap. This scenario has repeated itself over and over since the beginning of the industrial revolution, whether you look at textiles, steel manufacturing or agriculture. However, someday the cheap land and human resources will run out.

I realize that the problem of commodification of the winegrape is not an important issue for the wine communities in many states in the US because they are small and most of the wine sales are through tasting rooms. Furthermore, many growers in these areas grow winegrapes to make their own wine rather than sell them to a winery. Even though the dollar value of small wine communities is much lower than that of large wine communities in California, Washington, and New York, the financial success of both is still based on the fact that their wine is a value added product. Tourists come to the wine region, visit the tasting rooms, and see the scenery, and when they buy the wine they are buying an experience, whether it is Texas, Ohio, Michigan, or wherever.

What is the answer to the commodification of the winegrape? Ensure that wine remains a value-added product, no matter what the price point. This may sound odd because wine has always been a value-added product. However, the consolidation and globalization occurring in the industry, along with the rule changes that some are pushing for, is creating a more and more homogenous product that can be grown and made anywhere. Growers who sell their winegrapes to large wineries are being squeezed financially. The dilemma was really brought to my attention the other day when I was listening to a grower talk about the price of grapes. He said "We have to figure out a way to grow winegrapes cheaper because the wineries just don't care." It struck me that the long-term answer is not to figure out a way to farm more cheaply, because that is a losing battle. The long-term answer is to somehow figure out how to make the wineries care, to make them want your winegrapes.

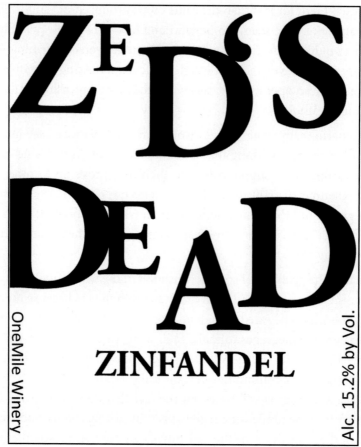

Figure 4-2   Selling wine in a way that makes it difficult to tell where it is from may contribute to winegrape commodification.

When one looks at the issues around commodification of winegrapes, the problem seems overwhelming. For example, how can a small grower in California who sells to a large winery with global holdings keep their grape prices from sliding even lower?

Trying to grow winegrapes more efficiently is always going to be part of the equation but it is not the solution all on its own. A grower needs to figure out how to continue to add value to their winegrapes so the winery wants them over someone else's. This could be accomplished by growing the best quality winegrapes possible, or by promoting the uniqueness of the region's wines, by using non-conventional farming practices such as sustainable, organic, or Biodynamic, or by enhancing your region's wine tourism. Once one has figured out how to add value to their winegrapes, they then need to market them aggressively. Any business that has something to sell has a marketing program that is always a significant part of its overall budget. Maybe growers should emphasize marketing their winegrapes to wineries to the same degree that wineries market their wine to consumers. Talk to anyone who has opened a winery in the last few years, and they will tell you that success is all about marketing, marketing, marketing.

If one looks at any agricultural product grown in the US that has suffered from global competition or commodification, one will see growers trying to figure out how to add value to their product to differentiate it from others as a way to survive. Since wine has always been a value-added product our challenge is to keep it that way. Higher winegrape prices ensure a sustainable business and provide the grower with more flexibility in terms of implementing more sustainable winegrowing practices, allowing them to move along the sustainable winegrowing continuum.

5

The Role of Science in
Winegrape Growing in the US

This may seem like an odd topic to many readers. Some would say "Of course winegrape growing is a science-driven endeavor." However, I often ask myself this question because I have observed things like the following: Federal and State support for University Cooperative Extension programs around the US eroding; the amount of research funding available through industry and government being woefully inadequate relative to the economic value of the wine industry; grower representatives from some wineries advising growers when to spray for leafhoppers or mites in their vineyards even though they do not have a shred of data on which to base their recommendations; and many wine writers waxing eloquent about Biodynamic farming when it is clear they really do not understand what it is. I continually read articles in trade journals or news media claiming there is only one true way to make quality wine and that way is whatever the bias of the author is. It can be Biodynamic, organic, dry farmed, made with wild yeast, done using only gravity flow, or whatever is the hot topic of the day.

Figure 5-1   University Cooperative Extension programs, like the one at the University of California, have experienced significant funding reductions over the last ten years.

One of the things I like about the wine industry is that anyone with interest, drive, and the ability to learn on the job can succeed in it. I have met many winegrape growers and winemakers who have had no or limited formal education in viticulture or enology yet grow really high quality winegrapes and/or make great wine. That said, not having a formal education in viticulture and some of the important biological sciences such as pest management, soils, and plant physiology can have its downsides. It is particularly obvious to me in relation to pest management, because that is my area of expertise. Moreover, the growing interest by growers and winemakers in sustainability has created a market for all sorts of new vineyard inputs such as soil inoculants, parasite and predator releases, foliar sprays such as compost teas, and so forth. Without a basic understanding of disciplines like plant physiology, soil science, and pest management it can be extremely difficult to make good decisions about what to do in the vineyard or what inputs to use.

What makes for a good educational background in preparation to become a winegrape grower? I will answer this question in part by relating a realization I came to a few years ago. In 2006 Washington State University announced that it was the first university in the US to offer an organic farming major. My first reaction was that it made sense because interest in organic farming has grown significantly in the public's consciousness and therefore it seems logical that the interest in doing it professionally would also grow, creating a demand for teaching it at colleges and universities. I then started to mull over in my mind what courses might be required for such a major and came up with the following: basic classes in botany, zoology, entomology, plant pathology, field ecology, chemistry, organic chemistry, physics, math, statistics, economics, geology, and soil science; and advanced classes in plant physiology, field ecology, invertebrate zoology, limnology, integrated pest management, biological control, and insect or animal population ecology. In other words, a very sound understanding of the biological and physical sciences, since ecological systems operate according to the principles studied in these disciplines, and understanding these principles should help one farm sustainably. I looked up the course requirements for the organic major at WSU and I was pretty close. Then it struck me that this is exactly the educational training I had as an undergraduate major in forestry from 1968 to 1972.

A point was made in the introduction to the organic major that organic agriculture was significantly different from conventional agriculture. That also seems to be the message that many practitioners and many writers are conveying. However, the fact that I had the same educational training majoring in forestry over 30 years ago as one experienced by an organic agriculture major in 2006 made me ask myself the following question: How different are the scientific underpinnings of conventional agriculture and organic agriculture? Hopefully by trying to answer this question, I can get back to discussing the role of science in winegrape growing.

In my opinion, regardless of whether someone chooses to farm conventionally, sustainably, organically, or Biodynamically, if they go to school to learn the basics of these farming approaches they should all have similar coursework. That is because understanding the principles of biology (including ecology) and the physical sciences is a great foundation for learning how to be a good farmer. My guess is that if one did a survey of agriculture majors at colleges and universities around the country, one would find that the coursework is pretty similar. Some people will say "That is the problem; ag schools teach conventional farming and don't teach organic farming principles!" Is that really true? All the basic biological and physical science classes are required for both majors. I think the problem lies in what happens to people once they leave school; in conventional agriculture the applying the principles learned in biological sciences tend to take a back seat to other things like risk aversion, reluctance to experiment with more sustainable farming practices, and concern that sustainable practices are uneconomic.

If one accepts the argument that all forms of agriculture are based on the same biological and physical scientific principles, then conventional agriculture can be viewed as farming with synthetic and some organic inputs, organic agriculture is farming with only organic inputs, and Biodynamic agriculture is where the farmer follows the tenets of Rudolf Steiner published in 1928 and as required by the Demeter certification program's farming standards. Some are going to say this is outrageously simplistic, but is it?

I now want to bring into the discussion two sciences that I have not mentioned yet and which I think will help us get at the core of some of our problems in how we talk about and view different approaches to farming: they are psychology and human behavior. People practicing or studying organic farming, sustainable farming, Biodynamic farming or conventional farming have pretty much divided into "us vs. them" camps. Each one thinks the other is wrong and way off target, feeling their views are justified due to their better understanding of the science underpinning the farming practices they use and promote. It is human nature to think that whatever type of farming one supports must be the right one. However, I think the division into groups supporting the various farming paradigms is more a result of psychology and human behavior than because one is working with a different, dare I say better, set of scientific principles. Some will ask "But what about the environmental degradation that has occurred as a result of conventional farming?" My answer is that these serious environmental impacts are due to implementation of the wrong practices as a result of an inadequate understanding of how they operate under fundamental biological and physical principles, not because conventional agriculture is based on a different set of scientific principles from those of the other farming approaches. Moreover, only a certain set of these practices are the culprits, not the entire farming approach. My guess is that if everyone switched overnight to organic farming, serious environmental problems would arise just as they have in conventional agriculture and for the same reasons: we [humans] have an inadequate understanding of how certain practices will affect the environment.

So what is the role of science in winegrape growing? In my opinion its application is essential for the future success of the US wine industry. Winemaking is often talked of as an art. That may be so, but I cannot evaluate this statement because I do not know winemaking. However, I do know something about winegrape growing and my scientific training and experience tells me that to be sustainable in the global wine market we cannot treat grape growing as an art; we must approach it scientifically. Moreover, we must try our best to recognize when psychology and human behavior are interfering with the science of growing winegrapes. I will give a quick example that I mentioned at the beginning of this chapter to illustrate my point.

THE IMPORTANCE OF OBJECTIVITY

Ideally, using a science-based approach to growing winegrapes ensures that one uses objectivity to make decisions. What do I mean by objectivity? An online dictionary defined "objectivity" as "judgment based on observable phenomena and uninfluenced by emotions or personal prejudices." However, I realize we are all human and it is probably impossible to be totally objective. We all have our biases no matter how objective we might think we are. Nevertheless, I feel it is worth being as objective as possible in sustainable winegrowing. In the following paragraphs I will present some real-life examples that might help convince you of the importance of objectivity.

The first case illustrates the importance of being an objective listener. I participated in a grower seminar where a winery had made red wine lots from two different vineyards; one which had experienced mite damage, and the other which had not. The wine from the mite damaged vineyard had less color and a less desirable flavor compared to the wine from the vineyard where no mite damage had occurred. All those in attendance were very interested in the tasting, because while most growers and pest managers assume that mite damage delays ripening and adversely affects wine quality, there is very little hard data to demonstrate it. This experiment seemed to add credibility to these assumptions.

Trying to be objective, I asked the presenter how much damage had occurred in the mite-damaged vineyard. The answer was "I don't know; it wasn't measured." In my mind this went from being an interesting project to a totally useless one because the important thing is not that mite damage affects wine quality, but knowing how much mite damage is needed to adversely affect it. I am afraid the lack of sufficient data collection in this study, which would have provided the needed objectivity, could lead to unnecessary spraying for mites. Why? Because with the winery looking at the vineyard over the grower's shoulder and having done experimental wine lots that showed mite damaged vines produce poorer quality wine, who can blame a grower for treating the vineyard when just a few mites show up?

The second example illustrates the importance of being an objective reader. In 2004 J. K. Reilly, writing for *Fortune* magazine, reported on a blind tasting pairing and comparing "biodynamic vs. conventionally made wines" (Reily 2004). I assumed this meant the grapes were from vineyards farmed either Biodynamically or conventionally. Twenty wines were paired based on "proximity of vineyard sites and with consideration given to price range. The wines in each pair were from identical or comparable vintages." In only one of the 10 pairs was the conventionally-made wine judged superior to its Biodynamic counterpart. Supporters of Biodynamic farming may find these results heartening, but let's put our objectivity hats on. We were not told how the grapes were grown in any of the vineyards other than that one set was called conventional and the other Biodynamic. The article contained nothing about the characteristics of the vineyards such as soil type, root stock, clone, irrigation regime, canopy management, and so forth. Objectively speaking we cannot attribute the results to the superiority of Biodynamic farming unless we know much more about the vineyards. To the author's credit they attributed the results more to the observation that Biodynamic growers are by nature concerned with the very fine details of farming, and if their wines are better it is as likely due to this as to Rudolf Steiner's tenets of Biodynamic farming. However, the downside of publishing the article is illustrated by the fact that I have seen it cited to support the idea that Biodynamic farming produces superior wines.

One sure way to increase objectivity in presentations or reports is to base them on studies designed for and analyzed using statistical tests. However, even when statistical analyses are used there are pitfalls in reporting the results. So it is important to be an objective reader or listener even in these situations.

The statistical tests are very valuable tools. However, like any tool they can be misused or not used at all. I had the fortune of taking several statistics classes from Dr. Leonard Marascuilo, a specialist in analysis of variance (ANOVA) and multivariate analysis at the University of California Berkeley. In fact I felt Dr. Marascuilo was the best teacher I had in all of my undergraduate and graduate experience. One class I took from him was devoted entirely to ANOVA. Some of the many things I learned were the proper application and interpretation of this powerful statistical tool. The following discussion is drawn from his lectures and excellent book on statistical methods (Marascuilo 1971).

How many of you have read a report or heard a presenter say the results of their study were "highly" significant? Or read or heard them say their results were significant to the $p \leq .001$ level or some other very small number? My guess is all of you have. Did you realize there is no such thing as a "highly significant result" from a statistical test? The results are either statistically significant or not. Statistical tests are based on probability theory and one cannot use an adjective in front of the word "significant" just to make the conclusions sound more convincing. Furthermore, when someone says their results are significant to the $p \leq .001$ level, or an even smaller number, it sounds great, but they are reporting the results of their statistical tests incorrectly. I will present a quick statistics 101 discussion to explain why this is the case and why it is important.

Let's say we want to test the efficacy of two new miticides and our test site is a 30 acre vineyard infested with Willamette mites. We will analyze the results using analysis of variance (ANOVA), on which most common statistical tests are based. The goal of our study is to determine if the miticides are efficacious, in other words do they work. To meet the requirements of ANOVA we will divide the vineyard up into 30 equal pieces called plots and randomly assign our "treatments" to these pieces, the treatments being no miticide (the control), miticide A, and miticide B. Ten pieces will be control plots (i.e., treated only with water), 10 will be treated with miticide A, and 10 will be treated with miticide B. Statistically speaking each plot is called a replicate, or rep, so there are ten reps for each of our three treatments.

Figure 5-2   Pacific mite infestation on a grape vine.

Once we have done our spraying we will go into each plot and estimate the percentage of leaves infested with live mites, giving us ten samples for each treatment. After we collect these data we are ready to do the ANOVA. An ANOVA calculates something called an F value which is used to look up in a statistical table to see whether it is statistically significant or not. It is standard practice to select the "p" or probability value of less than or equal to .05 (i.e., 5%) as the level for statistical significance. Therefore, we look up the F value for $p \leq .05$ in the statistical table and if the F value from our ANOVA is larger than the F value in the table, then the results of our study are statistically significant, i.e., at least one of the miticides worked better than water.

Unfortunately, what some researchers do is go to the statistical table and look up the "p" value that corresponds to the F value from their ANVOA and report their results were significant at that p value. If the F value is high then the p value will be low, e.g., $p \leq .001$ or even $p \leq .0001$. Since the p is so small they naturally conclude their results were "highly" significant. However, this is not the correct way to interpret the results from the ANOVA. All you can say is that your test was significant at the $p \leq .05$ level if that is what you chose before the study. You cannot spice it up with adjectives like "highly" and report the p value your calculated F value corresponds to. The reason why is simple.

An F value is used to determine if the ANOVA result is statistically significant, and the p value is the probability of making what statisticians call a Type I error. A Type I error is rejecting a true statistical hypothesis. In the case of our study, our hypothesis is that the miticides are no more effective than water. Rejecting a true hypothesis would be that we conclude the miticides are effective, i.e. at least one of them was more effective that water, when in fact none of them are more effective than water. This would happen if our ANOVA produced an F value that was statistically significant but in reality none of the miticides were more effective than water. This happens if the data we collected and analyzed were not reflective of the real world situation we were testing. We want to avoid this, of course, so we set the p value as low as we can to minimize the risk. As mentioned above, the accepted level for most statisticians is $p < .05$. The reason you cannot pick the p value out of the statistical table based on the F value from your ANOVA is because once you have collected your data you have either made a Type I error or not, regardless of what p value your F value corresponds with in the table.

ANOVA only tells you if one or more of your treatments are significantly different from the others. One needs to do a "post hoc" or after the fact analysis to find out which ones are different and this is commonly done. Nevertheless, one still cannot say their results are "highly" significant with these post hoc tests for the same reason presented in the previous paragraph.

There is a way to get more meaningful information out of ANOVA result, but it is rarely used. It is a calculation called "explained variance" and is easily obtained from the results of any ANOVA analysis. Let's briefly return to our miticide efficacy trial for an example. When we measure the number of mite-infested leaves in each rep for a particular treatment (e.g. miticide A, B or control) odds are we would not get the same result in each one. For example, in one replicate for miticide A we might find 10% of the leaves with live mites and no live mites in another replicate. In other words our data within each treatment are variable. Explained variance is the measure of the amount of variability in the data "explained" by the treatment effect, in our case the miticides.

Let's say that we did our analysis, found that the means of our treatments were significantly different at the $p \leq .05$ level and that the explained variance of the data was 80%. What this means is that 80% of the variation that occurred in our data was explained by the treatments we applied, i.e., the miticides. This is quite good. It also means that 20% of the variation in the data was due to things we did not measure, in other words unexplained variation. In my view explained variance is where the rubber hits the road with ANOVA based tests. It turns out that one can do a study that results in statistical significance but the explained variance can be as low as 20 to 30%. This means that 70 to 80% of the variation in the data is due to completely unknown factors. In any study there will always be unexplained variation but when it is small we know the study was well designed and the results biologically meaningful. Unfortunately, many researchers do not report explained variance. They only report statistical significance, which is only part of the story.

I have heard some proponents of sustainable agriculture state that "science" is what got us into the situation in which we find ourselves, with a degraded environment, hungry people around the world, and the domination by "industrial" agriculture. I find this a very scary statement. I will admit that I have totally bought into the scientific method as being the best way forward. I do not see any other viable approach to dealing with the problems in agriculture. Scientific objectivity and understanding clearly can lead to unintended consequences but science and objectivity should then be used to solve those problems. Good intentions just are not enough. We also have to accept that while we endeavor to be as scientific and objective as possible, several other things also influence what is done in the end, such as human behavior and politics.

One final practical comment: if someone recommends a sustainable winegrowing practice that has not been demonstrated efficacious through objective experimentation, be skeptical. One situation that I find particularly frustrating is when I encounter a grower using a practice that is considered environmentally "friendly," but its efficacy has not been demonstrated. It is as if the grower assumes that because a practice is environmentally friendly its efficacy does not need to be verified. However, this is not a sustainable approach. Any practice that takes either time, energy or both to implement should not be used unless it has been shown to be economically viable along with its environmental soundness and social equity.

REFERENCES

Marascuilo, L. A. 1971. *Statistical Methods for Behavioral Science Research.* McGraw Hill, NY. 578pp.

Reily, J. K. 2004. "Moonshine, Part 2." *Fortune* 150(4): 34–35.

Part II

Practicing Sustainable Winegrowing: An Holistic View

6

A Holistic Vision for the Farm: The Foundation of Sustainable Winegrowing

The importance of creating a holistic vision for your farm is captured in the following quote attributed to the famous New York Yankee catcher and erstwhile philosopher Yogi Berra: "If you don't know where you are going, you might end up some place else." If you have never taken the time to sit down and think about where you want to go with your winegrowing business, you may suffer the fate Yogi so eloquently describes.

During the process of creating the *Lodi Rules for Sustainable Winegrowing* farming standards (*www.lodirules.com*) and writing the second edition of the *Lodi Winegrower's Workbook* (Ohmart et al. 2008), I grew to appreciate that developing a holistic vision for one's farm should be the first step in implementing sustainable winegrowing. I arrived at this position through working on both projects with Kent Reeves, at that time a wildlife biologist for East Bay Municipal Utility District. Kent has training and experience helping producers, particularly cattle ranchers, develop holistic visions for their farms. Your holistic vision provides the foundation for your sustainable winegrowing program.

WHAT IS A HOLISTIC VISION?

Sustainable winegrowing is more than a laundry list of practices one uses in the vineyard. It is having a vision to ensure the long-term health (economic, environmental, and social), biodiversity, and productivity of the farm. Once you create a vision for your farm, each practice applied to the vineyard can be evaluated as to whether it moves you toward or away from your vision. Invariably some practices will take you away from your vision, because farming always involves some compromises. However, the fact that a specific practice does not move you toward your vision does not necessarily mean it should not be done. Just knowing where it fits into the holistic vision for the farm is important for achieving the goals of the vision. Furthermore, if a practice does take you away from your vision, you are not doing so in ignorance. Ideally, a holistic vision is developed collectively by key people in the farming operation and then shared with other family members and/or employees.

It is difficult for anyone not having experienced the vision development process to appreciate its importance. I will try to convey this by describing how it has affected the farming operations of Bokisch Vineyards & Winery in Lodi, California (Ohmart 2008; www.bokischranches.com).

By developing a holistic vision for his farming operation, Markus Bokisch realized that he held the keys to creating buy-in from all the parties connected to his farming operation, from his family and employees to the financial partners in some of his jointly-owned vineyards. He already had sustainability goals prior to joining the *Lodi Rules* program. However, the vision and the process he went through to create it as part of the program helped him to bring all of the people involved in his farming operation together to achieve those goals.

For example, by sharing his vision with the financial partners in some of his vineyards he was able to convince them of the importance of investing in improving wildlife habitat by putting up owl boxes, bat boxes, songbird boxes, and wood duck boxes, and establishing hedgerows around vineyards and planting more oak trees. Furthermore, by sharing his vision with his vineyard workers, he helped them appreciate their vital role in producing the highest quality winegrapes possible. He can trace the effects of the vision directly to the quality of the wine produced from the certified vineyards because the workers have taken added pride in the work they do, from leaf removal to shoot-thinning, knowing they are major contributors to the quality of the wine.

The best way to develop a holistic vision for your farm, in my experience, is to attend a facilitated workshop where you are taken through a series of steps to create a vision. That is because the process one goes through to create the vision is so important. Creating a holistic vision includes defining your resource base to be managed and developing one or more sustainable goals. For each goal, you develop objectives, strategies to achieve them, and a monitoring scheme to determine if the goal has been met. This is easier said than done, however, because the only place I know where this is possible in the wine world is in Lodi, California, where the Lodi

Figure 6-1   Song bird box in a tree adjacent to a vineyard.

Winegrape Commission offers these workshops for growers who are joining the *Lodi Rules for Sustainable Winegrowing* certification program.

As a part of the *Lodi Rules* program we developed written guidelines for developing a holistic vision. Since it is such an important part of sustainable winegrowing, I will present the guidelines here for those that may want to try this at home, so to speak. The examples provided come from a series of sustainable vision workshops with about 80 Lodi growers over a period of about six years. The resulting vision is only an example of what can come out of a visioning workshop and should not be taken as "the vision" for all growers in the Lodi region or anywhere else.

DEVELOPING A HOLISTIC VISION FOR YOUR FARM

Developing a Holistic or Sustainable Vision for your farm involves identifying your farm's resources, creating a set of sustainable goals, and then producing a sustainable vision for your farm. This process will help ensure the long-term health, biodiversity, and productivity of your farm and the surrounding ecosystem.[1]

IDENTIFYING AND DEFINING THE YOUR FARM'S RESOURCES

The resource base for your farm includes not only the land you farm, but also the tools and people available to help you in that endeavor. The following guidelines will help you identify your farm's resources:

LAND BASE

Describe and map the physical boundaries of your property that you manage. This should include: total acres, neighbors, driveways and road systems, water access rights, streams and riparian corridors, vernal pools, wet swales, drainages, degree of slope, any existing erosion, and the presence of animal and plant species including any threatened or endangered species that may affect farming on the site and/or be subjected to local, state, or federal regulations.

Note: The examples used to illustrate the various components for developing a holistic vision came from a workshop where a group of Lodi Winegrape Commission members created a Holistic Vision for the entire Crush District #11.

Example Land Base

The vineyards managed by Lodi Winegrape Commission (LWC) members within the land delineated as Crush District #11 (Figure 6-2). (Your site description will probably be more detailed and only include the property on the farm for which this holistic vision is being created)

PARTICIPANTS

List all the participants who will be involved with the operations. These are individuals needed to make the ecosystem and sustainability management solutions viable. This may include any or all of the following: family, friends, employees, neighbors, community members, other stakeholders, bankers, farm advisors, pest

---

1. *www.lodiwine.com/The%20LodiRulesCompanionDocumentVersionR-LW-S-6132.pdf*

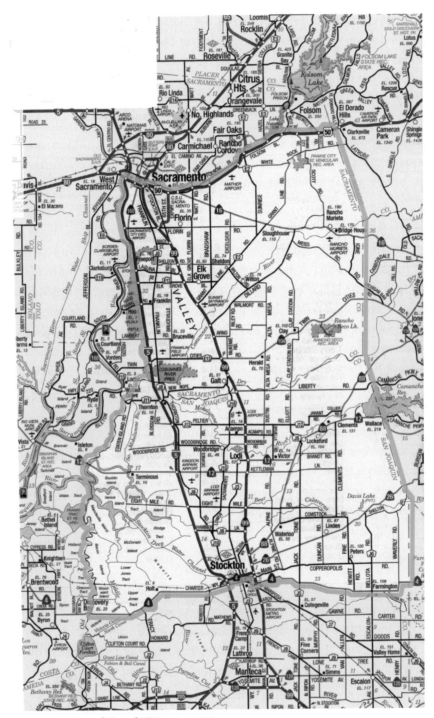

Figure 6-2   Map of Crush District #11.

management consultants, public officials, regulators, environmental groups. Since ecosystems do not stop at traditional property boundary lines, a good idea is to look across boundaries and develop an active partnership with neighbors and other stakeholders. Try to be as inclusive as possible.

Example Participants

Winegrape growers, winery owners, suppliers, retailers, consumers, politicians, government agencies, educators, crop advisors, neighbors, laborers, financial institutions, foundations, media, and non-government organizations, LWC staff.

AVAILABLE RESOURCES

Brainstorm about what resources you have available to manage your farm. These include financial resources, human resources, capabilities of the participants, and equipment. List the major physical resources from which revenue is generated and the people who influence or will be influenced by management decisions. List the capabilities of people who can act as resources for your operation. (There may be a slight blurring here between *Participants* and people who act as a *Resource Base*; it is okay for there to be an overlap.)

Example Available Resources

Operating budget for the LWC for one year and the capabilities of the *Participants* and people involved who influence Crush District #11. The capabilities of the people who influence the LWC include cooperation; leadership; education; experience; time; research involvement; compassion; energy; understanding; risk takers; knowledge of opportunity; and commitment.

DEVELOPING SUSTAINABLE GOALS

Now that you have identified your farm's resources to be managed, it is time to identify some possible sustainable goals for your farm and objectives to meet those goals. There are three things to consider in developing your sustainable goals: the shared values among the farm's participants; the forms of production you wish to achieve from the farm; and the desired future landscape of your farm.

The following are guidelines relating to each.

SHARED VALUES (QUALITY OF LIFE)

This is a statement articulated by the participants of the sustainable management process for your farm (family, employees, stakeholders, etc.) and is the foundation of the vision. It is an expression of the way people want to live their lives and work together based upon what they value most. What are your social values? What draws you to the farming life? What values do you want reflected in your farm? How do these values influence the overall quality of life of those participating?

Example Shared Values

A brainstorming by the workshop participants revealed these shared values: love, security, family, health, purpose, companionship, happiness, peacefulness, peace, stress relief, time, harmony, a reason for all this, intrinsic worth, to feel good, and fun.

Some readers right about now might be thinking that this process may not be for them for various reasons, including not being interested or ready to become involved in sustainable winegrowing. However, my experience based on doing more than 10 of these workshops with Kent, attended by almost 200 growers with the complete range of views about sustainable farming, is that everyone who participates in the workshop becomes invested in the sustainable winegrowing process.

The following statements are a summary of what the above brainstorming indicates about the participants' shared values:

- We value a life surrounded by family and friends filled with companionship.
- We feel good and secure in our homes and community.
- We want our lives to be filled with love, purpose, happiness, harmony, and peace. And, because we strive to fill our lives in a positive way, there is less chance for stress.
- We would like time for fun and to enjoy the intrinsic worth of our lives with friends and family.

FORMS OF PRODUCTION

From the land base being managed, what supports your shared values or quality of life? What do you need to produce to achieve your desired quality of life? Examples include: profit, recreation, culture, aesthetics, other products, etc.

Example Forms of Production

A brainstorming by the workshop participants revealed these forms of production:

- Produce a profit from winegrapes
- Produce good quality winegrapes and wine
- Produce name recognition
- Produce thirsty buying customers
- Produce content, viable, and skilled workforce
- Produce demand for Lodi wine
- Produce clean air, water and healthy soil
- Produce good community relations
- Produce a good environment

The following statements are a summary of what the above brainstorming indicates about the participant's shared values:

- In order to sustain our *Shared Values,* we will generate profit from high quality winegrapes that produce high quality wine.
- This will create name recognition and demand for Lodi wine from thirsty buying customers.
- We will produce good relations with the local community.
- Our workforce will be content, viable, and skilled.
- Our winegrape production will result in clean air and water, healthy soil, and a robust environment.

DESIRED FUTURE LANDSCAPE

Describe what the landscape of your farm would look like in the future in order to sustain production and quality of life. Where would you like the farm to be in five years? Ten years? When you retire and pass the farm along to your children? You may include components of the land base that you described above. Historical knowledge of what the landscape used to look like and what has been done to change it may help you develop this landscape description. You may include pictures, maps and/or a written description of your desired landscape.

Example Desired Future Landscape

For our forms of production to be truly sustainable, our future land base will have a healthy and diverse soil with good populations of soil microbes. There will be an abundance of available nutrients for all plants and we will create an appreciation within the community for the importance of a healthy soil.

The water cycle will be adequately beneficial to crops, land, people, and nature. There will be an abundance of water from healthy rivers, streams, other surface water sources, and ground water. There will be greater availability and affordability of water with a balance of use between urban and rural communities.

Sufficient solar energy will be captured efficiently by healthy grapevines and native plants. Native vegetation communities will be thriving and biologically diverse. Renewable energy sources (e.g. solar power) will be used by more people and businesses in the community. This will create more sources of sustainable energy. These practices will result in better air quality for our community.

There will be adequate healthy and diverse biological communities. These communities will have ample vegetation of native plants that in combination with abiotic components provide good habitat for healthy wildlife populations. These

Figure 6-3   Irrigation canal planted with native vegetation that provides habitat for wildlife and stabilizes the canal banks while not impeding water flow.

environments will also provide for diverse insect and other invertebrate populations beneficial to agriculture. Our community will understand and recognize the
importance of a diverse and robust ecosystem.

There will be good rural and urban cooperation in the pursuit of our shared
vision. This will provide a balance of good agricultural land and sustainable economic growth. Urban centers are planned and clustered on non-arable land preserving our small town and farming atmosphere. Our community will value
vineyards, sustainable agriculture, and farmers.

Putting It All Together

You now have all of the information you need in order to create a holistic vision for
your farm. Basically you put together the description of the whole that you will
manage, which consists of the physical boundaries of your farm, the participants in
your vision, and the resource base. Then you add the three-part vision which is
quality of life you desire, the forms of production from the farm, and the future
desired landscape. Below is the shared Holistic Vision for the growers who
attended the LWC workshop, where the above examples are taken to illustrate what
one looks like in complete form after going through the above exercises. It is
important to note that this is not "the" vision for LWC but is the shared vision for
the growers who attended the workshop.

AN EXAMPLE HOLISTIC (SUSTAINABLE) VISION FOR THE LODI WINEGRAPE COMMISSION

WHOLE TO BE MANAGED

Physical Boundaries—The land delineated as Crush District 11 and the vineyards managed by Lodi-Woodbridge Winegrape Commission members.

Decision Makers or the People Involved Who Influence the LWC—winegrape
growers, winery owners, suppliers, retailers, consumers, politicians, government agencies, educators, advisors, neighbors, laborers, financial institutions,
foundations, media, and non-government organizations.

Resource Base—Operating budget for the LWC and the capabilities of the decision makers and people involved listed in the previous paragraph who Influ-

ence Crush District 11. The capabilities of the people who influence the LWC include: cooperation, leadership, education, experience, time, research involvement, compassion, energy, understanding, risk takers, knowledge of opportunity, and commitment.

THREE PART VISION

Quality of Life—We value a life surrounded by family filled with companionship from them along with our friends. We feel good and secure in our homes and community. Our lives are filled with love, purpose, happiness, harmony, and peace. And because we strive to fill our lives in a positive way there is less chance for stress to creep in. We have time for fun and to enjoy the intrinsic worth of our lives with family and friends.

Forms of Production—In order to sustain our Quality of Life, we will generate a profit from good quality winegrapes that produce good quality wine. This will create name recognition and demand for Lodi wine from thirsty buying customers. We will produce good relations with the local community. Our workforce will be content, viable, and skilled. Our winegrape production will result in clean air and water, healthy soil, and an overall good environment.

Future Resource Base—For our Forms of Production to be truly sustainable our future land base will have a healthy and diverse soil with good populations of soil microbes. There will be an abundance of available nutrients for all plants and we will create an appreciation within the community for the importance of a healthy soil.

The water cycle will be healthy and beneficial to crops, land, people, and nature. There will be an abundance of water from healthy rivers, streams, other surface water sources, and ground water. There will be greater availability and affordability of water with a balance of use between urban and rural communities.

Sufficient solar energy will be captured efficiently by healthy vigorous vines and native plants. Native vegetation communities will be thriving and biologically diverse. Renewable energy sources (e.g., solar) will be used by more people and businesses in the community. This will create more sources of sustainable energy. The air quality in our community will be good.

Figure 6-4   A community enjoying the benefits of a vibrant wine industry.
Courtesy of Lodi Winegrape Commission.

There will be adequate healthy and diverse biological communities. These communities will have ample vegetation of native plants that in combination with abiotic components provide good habitat for healthy wildlife populations. These healthy environments will also provide for diverse invertebrate populations beneficial to agriculture. Our community will understand and recognize the importance of a diverse, robust, and strong ecosystem.

There will be good rural and urban cooperation within our diverse community with participation in the continued development and pursuit of a shared vision. This will provide a balance of land use with protection of good agricultural land, so it is available for use while encouraging sustainable economic growth. Urban centers are planned and clustered on non-arable land preserving our small town/farming atmosphere. Our community will value vineyards and sustainable agriculture.

CREATING A PLAN TO ACHIEVE YOUR HOLISTIC VISION

Based on your answers above, brainstorm about actions that you will need to take in order to produce your desired future landscape vision. Determine whether each of these actions is shorter term (1-5 years) or longer term (6+ years). Then, list the goals, objectives, strategies, and monitoring programs that you can use to accomplish these actions.

Goals should be longer term. Objectives are shorter term and when combined with other objectives, achieve the longer term goals. Strategies are the actions used to produce the objectives. Monitoring evaluates the effectiveness of your strategies to ensure that your goals are met.

EXAMPLE GOAL #1

When I was growing up, I used to see a lot of quail and other wildlife on the farm and now I rarely see any. My goal is to see more quail and other wildlife on my property.

Objective 1—In order to see more quail on my farm, I need to improve their available habitat. I will do this by planting and maintaining a hedgerow of quail brush (*Atriplex* spp.) along 100 feet of my western fence line.

Strategy—Proposed actions to achieve this objective include spraying my western fence line with an herbicide in January, digging 25 plant holes in February, installing a drip line after the holes are dug, buying 25 small quail brush plants and planting them in early March. I will give the plants a deep watering once a week during the first summer or more often if needed.

Monitoring—I will monitor the plants to see if they are receiving adequate water during the first summer. I will monitor the weed population in the hedgerow and remove troublesome competitive weeds when necessary.

Objective 2—To reintroduce quail onto my farm.

Strategy—Once the quail brush hedgerow is established (1-2 years), I will reintroduce 3 breeding pairs of quail in the early spring. I will also provide three covered ground nesting boxes to protect the quail from predators.

Monitoring—Monitoring will include walking the hedgerow weekly to determine if there have been any causalities and to count the number of birds. I will record the number of quail observed, the time and date on each outing. Once a month I will examine the nest boxes and clean or repair them if needed.

Objective 3—To provide secure nesting sites for waterfowl using my reservoir.

Strategy—I will place 5 wood duck boxes along the reservoir edge, maintain the boxes not less than three times a year and cooperate with the California Waterfowl Association.

Monitoring—Monitoring will be done as a part of the maintenance program. Not less than three times a year, I will check and maintain the boxes; I will record the date, time, signs that indicate the presence of occupancy, number of inhabitants, and condition of the box.

EXAMPLE GOAL #2

I want to earn more money from my farm.

Objective—To produce high quality winegrapes and receive a premium price for them.

Strategy—Proposed actions to achieve this objective include more precise and improved water management, nutrient management, and pest management, and a more active roll in marketing my grapes to smaller higher paying wineries.

Monitoring—Demand for my grapes, number of long-term contracts, increased value, reduced costs, and bottom-line improvements.

REFERENCES

Ohmart, C. P. 2008 "Lodi Rules Certified Wines Enter the Marketplace." *Practical Winery & Vineyard* 29(5):32–42.

Reeves, K. 2008. "Ecosystem Management." *In* Ohmart C. P., C. P. Storm and S. Matthiasson. 2008. *Lodi Winegrower's Workbook* 2nd Edition. Lodi-Woodbridge Winegrape Commission, Lodi CA. pp. 15–63.

7

Social Equity: The Third E of Sustainable Winegrowing

I began Chapter 1 with a discussion about the three challenges to overcome in developing a sustainable winegrowing program. The first was defining sustainable winegrowing. While many definitions of sustainable farming have been proposed, one point of agreement for most is the foundation consisting of the three E's of sustainability; winegrape growing that is economically viable, environmentally sound, and socially equitable. Discussions, essays, and presentations on sustainable winegrowing almost always deal with the first two E's in great detail. Social equity, the third E, while usually getting mentioned, is the E that gets the least amount of time devoted to it, however. Why is that? I think it is because it is the E that is the most challenging to address for many companies. While thinking about the title for this chapter, I was tempted to call it the third rail of sustainability, rather than the third E, because no one wants to touch the subject for fear of getting zapped. For readers not familiar with the third rail reference, a third rail is the metal cable or rail that runs along the side of electric train tracks and which carries the very high voltage that powers the train. It is certain electrocution to touch this rail when it is "live." Touching the third rail is a term that some use for a topic that no one wants to discuss for fear of terrible things happening if they do.

I am not implying the wine industry has ignored social equity issues. There are many vineyard and wine companies which have devoted much time and many resources to developing exemplary human resources programs. However, it is fair to say the third E has not received the same attention as the other two. My own inattention towards this topic is a great example of how the third E gets neglected, either consciously or unconsciously. I drafted this entire book and sent it out for peer review without including a chapter on social equity. Fortunately, I realized this amazing oversight in time to add this chapter, thereby providing a complete picture of sustainable winegrowing.

What is social equity? There are several contexts in which it can be defined. In terms of conservation Wikipedia states "Social Equity implies fair access to livelihood, education, and resources; full participation in the political and cultural life of the Community; and self-determination in meeting Fundamental Needs"[1]. Social equity is about people having fair access to the things mentioned above, and people are the resource that is the foundation of any company. Human Resources (HR) is the label that has been given to the people that make up a company, including the owners. Managing HR effectively is how social equity is achieved.

There are many reasons why HR issues are so challenging to the wine industry and may explain why HR is consciously or unconsciously avoided. Three stand out as exceptionally challenging. One is the cost of labor, which is, with the exception of land prices in some situations, by far the costliest part of growing winegrapes. Many growers are either barely meeting or just exceeding cost of production. So any HR issue that involves an increase in wages and/or benefits might be a financial make or break situation for a grower. Social equity is achieved by having access to livelihood, education, and resources. Wages and benefits provide this access, so they are critical to the third E of sustainability.

A second big HR challenge for many workers is obtaining US citizenship. Illegal aliens make up a significant part of the agriculture workforce, and no one seems to have figured out how to deal with this situation in a way that satisfies all sides of the political and social spectrums. The third challenge is that people can be very difficult to manage. They complain, they can be stubborn, they have egos, and unlike a troublesome piece of equipment, they cannot be simply and easily replaced with a new one. Despite these challenges we must recognize that human resources are the most important part of our business, so everything should be done to meet the HR challenges we face today to achieve social equity in our businesses.

The rest of the chapter is devoted to important HR issues. I will be the first to admit to not being an HR expert. However, I am human, I have held many jobs, and I have supervised many people over my career and feel I have learned some things in the process. Job-related human resources issues can be viewed as falling into two main areas: one that deals with job satisfaction and a person's feeling of worth in a company, and the other involves the more matter of fact planning and processes used in managing employees effectively.

---

1. *http://en.wikipedia.org/wiki/Social_equity*

JOB SATISFACTION

Job satisfaction has always played a key role in my achieving peak job performance, and this is likely true for many others. A business with employees operating at peak performance means a successful and sustainable business. How many of you know people who are dissatisfied with their jobs and, as a result, are not committed to working as hard as they are capable? It is a very common phenomenon. There are lots of reasons for being dissatisfied in one's job, but often it is related to not feeling valued by the company.

One way to start building job satisfaction in employees is to get them involved in the company's holistic vision. The last chapter emphasized that a holistic vision is the foundation of a sustainable winegrowing program. Creating a holistic vision for your vineyard or winery business and sharing it with all employees is a great way to bring employees together. Better yet, have the employees participate in creating the company vision, because it will help them see their role in making the vision a reality. Many jobs in the winery and vineyard are difficult and tedious. Yet they are absolutely critical for the company's vision to be realized. If the people doing these jobs can see they are important contributors to achieving the company's goals, they will likely have good job satisfaction no matter what tasks they are asked to perform.

There are many other things that affect job satisfaction. One of the most challenging to discuss and implement for many companies is increased wages and/or benefits. However, if we are honest with ourselves, most of us gain some if not a lot of job satisfaction from how much we are compensated for our work. It indicates to us our worth to the company in very direct terms.

There are many indirect ways to build job satisfaction, too. Promoting an open, fun, and productive work environment is one. Another is convening fun and imaginative team building events. A third is a formal system that recognizes employees for years of service, job performance, or commitment to helping achieve the company's vision. Most of these ideas are doable for most companies and it simply takes some time and imagination to make them a reality.

Developing an effective human resources management program, particularly if you work in a medium to large company, involves a lot of planning, designing, implementing, and evaluating practices. Much of the following information is drawn from the human resources management chapter Dr. Liz Thach, Sonoma State University, wrote for the *Lodi Winegrower's Workbook 2nd Edition* (2008).

Figure 7-1   There are many ways to recognize an employee's contributions to the company.

HUMAN RESOURCES MANAGEMENT

Human resource management can be divided into at least four major topic areas; 1) staffing and recruiting, 2) training and organizational development, 3) employee relations, 4) compensation and benefits.

STAFFING AND RECRUITING

Establishing a strategic staffing, recruiting, and retention plan ensures your company will have the correct number of employees and appropriate skills needed to achieve your holistic vision and business strategy. Not retaining good employees is costly. The current estimate for the cost of replacing an employee is 1.5 times their annual salary. This figure is arrived at by adding up the cost to recruit a new employee, downtime, potential overtime or temporary employee costs, costs of management time to interview new employees, cost to orient and train the new employee, and potential unemployment costs. Staffing your company with talented employees, selecting the most effective recruiting strategies, and implementing practices to retain employees over the long term not only enables you to implement your business strategy effectively, it also will provide you with a competitive advantage in the global marketplace.

Good recruiting involves job descriptions for each position, a standard interview process, and utilizing a variety of methods to advertise job openings, such as by word of mouth, in newspapers, web recruiting, at job fairs, hiring a recruitment firm, and listing job openings in trade journals. If a range of methods is used it is important to track the results of each one to calculate their cost/benefit ratios to refine future recruitments.

A job description for each position has multiple benefits. It serves as a guide for developing the interview process for that job. It provides a guide for the person who occupies the position so they know clearly what is required of them, and it can be linked to periodic salary review and performance management. Given the increasingly litigious nature of our society, the job description should contain an Americans with Disability Act (ADA) section.[2]

It is good to have a standard interviewing process so that each candidate interviewed is considered using the same criteria as all the others. Interview questions should only be related to the job and generally fall into the following categories: prior work experience that may be relevant to the position for which you are recruiting; general skills and aptitudes related to the job criteria; education; attitudes and personality characteristics; and career goals and occupational objectives. It is very important to avoid illegal questions such as those appearing in Table 7-1 (Thach 2008).

Once you have gone through the hard work and expense of recruiting and hiring a talented person, retaining that person is your next goal. As mentioned above, job satisfaction is clearly one way to achieve this goal. It starts by making the person feel welcome and secure in their new position. One way to accomplish this is by having an established orientation program for new employees that includes a motivational overview of the company that could include discussing the company's holistic vision, an overview of employee benefits, work performance standards, and important company policies. It would also be good to have them meet with other key employees and tour the company's facilities and vineyards.

In the event that an employee leaves, it is very important to conduct an exit interview with them to try to determine why they left. This could be awkward if the employee left because they were unhappy. However, if there are conditions in the workplace that are causing talented people to leave, it is important to understand what they are so they can be rectified. More than once I have been exposed to workplaces where talented people came and went due to unacceptable working

---

2. *www.onestophr.com* is a good resource for language to include in the ADA section of the job description.

Table 7-1   Illegal interview questions.

1. Are you a US citizen?

2. Where were you, your parents, or spouse born?

3. Can you provide me with a copy of your birth certificate, naturalization or baptismal records?

4. What is your maiden name?

5. How old are you?

6. When did you graduate from high school or college?

7. What is your birthday?

8. What is your marital status?

9. How many children do you have and what are their ages?

10. Do you plan to have a family? If so when?

11. Do you have child care arrangements for your children while you work?

12. What religious holidays to you observe?

conditions, often related to poor management style, and the people in charge had no idea why employee retention was low. This is clearly not a sustainable company.

TRAINING AND ORGANIZATIONAL DEVELOPMENT

Good training and organizational development not only ensures your employees have the skills needed to accomplish their work, but it will also increase their job satisfaction, which has been proven to enhance job performance. It is important to remember that you are also a member of the company's workforce and therefore you should consider continuing education and training for yourself as well. The American Society of Training and Development did a 3 year study that verified that companies who invest in training report higher profit margins and higher income per employee. They also found that companies that were the most proactive about employee training from all industries invested an annual average of $955 per employee for training and encouraged every employee in their company to participate in some type of work-related training and/or development each year (Thach 2008).

Figure 7-2   There are many trade journals one can read to keep up on the latest information in vineyard management and the rest of the wine industry.

New practices are constantly being developed for improving wine grape quality and other aspects of sustainable winegrowing, so it is important to keep informed by reading trade journals, attending seminars, and networking with growers in the region, around the state and even in other regions of the US. Encouraging your employees to grow professionally through education and training will keep them stimulated and performing at an optimum level as well. The type, level, and amount of education and training will vary based on the type of job an employee occupies. For some key employees, consideration could be given to developing a formal career planning process and holding periodic discussions with them regarding their career goals and how they might be achieved. Incorporating this into a yearly performance review process is one way to implement such an approach. A section could be added to their performance appraisal form in which the employee documents their career goals and you and the employee develop steps that need to be taken in order to achieve the goals.

A certain level of safety training is a legal requirement, the details of which may vary from state to state. It is important to ensure you are up to date on the latest safety training laws and regulations. It is also good business sense to emphasize safety training. In California it is a legal requirement to conduct safety training for an employee any time they change jobs where the new job duties involve a hazard. It is a good idea to track safety statistics, such as time lost due to accidents, to evaluate the effectiveness of the safety training. You also might want to consider recognition for employees who maintain a good safety record.

If you have employees that do not speak or comprehend English very well, it is highly recommended that training also be conducted in their primary language or that a translator is present at the trainings.

Employee Relations

Establishing clear and positive employee relations policies and practices creates a more positive company culture and also decreases your exposure to legal liability issues. Moreover, it will create an open and more relaxed working environment. Employee relations is a two-way street. You as the employer want to create a working environment that is appreciated by the employee but you also want to create the situation where the employee has the desire to be a better employee for you.

An employee handbook, ideally in the primary language of the employee, is a very useful document. The format can vary according to the size of the company and the number of employees. The employee handbook for a small vineyard operation may simply be a few pieces of paper stapled together. For a large company it may be a bound handbook or a digital book that the employee keeps on their computer or has access to via the Internet. The handbook contents likely will include the company's work standards, policies, and overview of employee benefits. It is important that the employee handbook is reviewed on a regular basis, maybe annually, and updated when necessary. It could also contain a description of the job performance system that describes how the employees will be evaluated for job performance, as well as a grievance process an employee would use to bring forward a complaint or concern and detailing how it would be investigated. It is important that these processes be perceived by the employee as fair and working properly.

It is important to have regular employee meetings to encourage good and open communication. The significance of a recognition program for employee job performance, years of service, and safety record has already been mentioned.

COMPENSATION AND BENEFITS

Salary is the first thing most of us think of when employee compensation is discussed. Labor is the costliest input for growing winegrapes and in the extremely competitive global marketplace, increasing employee pay may be very difficult of not impossible. There is a range of possibilities for financial compensation that does not involve a salary increase. Some are providing a pension plan, a pre-tax flexible spending account for health, child care, or elder care, or a company matching 401k plan. Health benefits and paid time off are other forms of financial compensation. There are also forms of financial compensation related to job performance such as paying piecework rates, merit pay for exceptional job performance, a program that gives an employee a bonus based on improvements in productivity, profit sharing, and stock options.

The purpose of an employee recognition system is to recognize employees for contributing to the overall business strategy of the company and its holistic vision. Contributions come in many forms such as good work ethics, good safety performance, exceptional customer service, length of service, teamwork, or community service. Recognition might include praise, a gift certificate, a team outing, a bonus, or a salary increase and/or promotion.

Many companies with progressive approaches to HR have found that it is best to encourage employees to use vacation and holiday time off rather than have it accrue. This not only helps to insure a good balance between work and non-work life but also reduces the company's administrative burden of tracking unused vacation and having to make large payments when an employee leaves. Another innovative practice is to reward employees with a bonus at the end of the year for not getting sick and using sick time. Some companies are pooling all paid time off from vacation, holiday, sick, and personal leave into one account which allows the employee to use the days in a way that suits them best as long as it does not interfere with critical work schedules (Thach 2008).

There are many other ways to compensate employees and some good books have been written on the topic (e.g., Aguanno 2004; Nelson et al. 2005; Putzier 2001; Venice 2009).

SUCCESSION PLANNING AND ESTATE PLANNING

Planning for passing the company from one generation to the next both in terms of leadership as well as financial assets can be a very challenging process. First, we all want to think we will live forever. Second, none of us wants to think about something happening to ourselves or other critical members of the company that prevents them from carrying out their duties effectively. Third, financial discussions can be very difficult. Often these decisions are left until something happens that forces the issue and then decisions are made under duress. A sustainable business needs to have succession and estate planning in place.

It is important to distinguish between succession and estate planning. Succession planning deals with passing company leadership from one person or persons to another. Estate planning involves transferring financial assets from one person or persons to another. Succession planning involves human and intellectual capital (i.e., people) while estate planning deals with financial capital.

Succession planning usually is done just for the top management team and occasionally for the next level below if the company is a large one. It involves identifying one or two successors for the company president and ideally entails development opportunities and/or mentoring for them to prepare them for the eventual succession. Keep in mind succession can become necessary for a variety of reasons. The current company president may want to retire, may become ill, or have a serious accident and not be able to carry out their duties effectively, or may die. While we all would like to think we are irreplaceable, I believe a sustainable business is one where succession occurs smoothly and the company continues to thrive afterwards.

Estate planning is probably best done with professional help from lawyers and/or accountants to ensure that when financial assets do get transferred from one generation to another for from one person to another, it occurs smoothly, in a timely manner, and with minimum cost to the company.

COMMUNITY INVOLVEMENT

Social equity also involves the community within which your vineyard operation exists and where you live. Being a good neighbor in your local community and an active community member contributes to its sustainability. The key to finding out what the important issues are to others in your community is to communicate with them. Two of the biggest barriers to problem solving are lack of communication and a persistence of misinformation. Take time to research local issues. Learn from other community members their concerns by attending meetings or workshops on local issues (Dlott et. al. 2002). Then if time will allow, get involved in community work that interests you the most. There is no shortage of work to be done on community housing, transportation and traffic, education, health care, water quality and supply, and smart growth.

Your vineyard operation also exists within another community, the wine community. Being an active member in this community is critical to its sustainability. It is likely a local grower and/or winery group exists where you live. There is almost certainly a statewide organization in the state where you live, and there is Wine America and the National Grape and Wine Initiative that are active at the national level in the areas of federal government policies related to wine and grapes as well as in research on marketing, viticulture, and wine making. These organizations survive on voluntary contributions of time and money by people who recognized that the sustainability of the wine community is enhanced by their involvement. Consider becoming involved in your local or state organizations. With time you may find you want to become active at the national level. The success of your vineyard operation depends in no small part on the sustainability of your local, state, and national wine community.

Social equity, the third E of sustainability, is complex and challenging to implement well. However, doing it well will likely mean the difference between a successful sustainable business compared to one that never reaches its full potential or even fails. I have just scratched the surface of this topic but hopefully provided some food for thought along the way.

REFERENCES

Aguanno, K. 2004. *101 Ways To Reward Team Members for $20 (or Less!).* Multi-Media Publications, Inc. 111pp.

Dlott, J., J. Garn, and Carla M. De Luca. "Neighbors and Communities" *In* Dlott J., C. P. Ohmart, J. Garn, K. Birdseye, K. Ross, eds. 2002. *The Code of Sustainable Winegrowing Practices Workbook.* Wine Institute & Calif. Assoc. Winegrape Growers. pp. 15.1–15.25

Nelson, B., K. Blanchard, and S. Schudlich. 2005. *1001 Ways To Reward Employees.* Workman Publishing Company.

Putzier, J. 2001. *Get Weird! 101 Ways To Make Your Company a Great Place To Work.* AMACOM, NY. 187pp.

Thach, L. 2008. *Human Resources Management. In* Ohmart, C. P., C. P. Storm, and S. K. Matthiasson. 2008. *Lodi Winegrower's Workbook 2nd Edition.* Lodi Winegrape Commission, pp. 268–295.

Ventrice, C. 2009. *Make Their Day. Employee Recognition That Works.* Berrett–Koehler Publishers, Inc.

8

Ecosystem Management: The Big Picture Approach to Sustainable Winegrowing

Chapter 6 helped you develop a holistic vision for your farm. Now it is time to focus on how the farming will be done and the practices implemented to achieve your holistic vision.

We farm within an ecosystem, so it is important to appreciate what it is and understand some of the basic cycles that operate within it. An ecosystem is the complex community of living organisms and their physical environment functioning as an ecological unit. It is important to recognize that the components of an ecosystem are inseparable and interrelated. You likely have heard the expression "the whole is greater than the sum of its parts." An ecosystem is a great example. It is made up of many components, and understanding how each one functions on its own is important but may not tell you how it functions when interacting with all the other components. It is important to look at your farm on an ecosystem level because then you will see how it operates as a unit. Moreover, an ecosystem management approach to growing winegrapes acknowledges that people are a part of and have a significant impact on ecosystem structures and processes, and that people depend on and must assume responsibility for the ecological, economic, and social systems where they live (Reeves 2008).

ECOSYSTEM PROCESSES

There are four fundamental ecosystem processes that determine its dynamics and the quality of the landscape: the mineral cycle, the water cycle, energy flow, and community dynamics. In order to look at your farm as an ecosystem it is important to know of these cycles and their basic functions.

MINERAL CYCLE

The mineral or nutrient cycle is the process by which key elements necessary for living organisms, such as nitrogen, phosphorous, and potassium, move through the living (biotic) and nonliving (abiotic) components of the ecosystem. Each element has its own cycle, which often is intertwined with the cycles of other elements. A healthy mineral cycle implies a biologically active soil with adequate aeration and energy flow below ground to sustain a variety of organisms in contact with carbon, nitrogen, and oxygen from the atmosphere. Implementing practices to prevent off-site nutrient losses (e.g., buffer strips, restored riparian areas, hedgerows, etc.) and increased nutrient cycling (e.g., cover cropping, adding compost, etc.) on your farm can help make it more efficient and environmentally sound (Reeves 2008). The Sustainable Soil Management chapter will discuss some ways a grower can affect the mineral cycle in their vineyards.

The carbon cycle is another important component of the mineral cycle. The recent alarming events, such as melting of glaciers and the polar ice cap, that appear to be a result of global climate change and which are hypothesized to be due to the dramatic increases in atmospheric CO_2 indicate the importance of the carbon cycle. The hypothesis is that CO_2, and other gases in the atmosphere are trapping heat from the sun in the atmosphere as it reflects off the earth's surface and tries to dissipate back into outer space, similar to how a greenhouse heats up on the inside. This is called the greenhouse gas effect and gases, primarily CO_2, are called greenhouse gases (GHG). There is still a lot to be learned about the role of vineyards in the carbon cycle, but its importance in the current discussions about the role of agriculture in greenhouse gas production warrants a brief mention here.

Here is a very simple presentation of the carbon cycle with emphasis on the word "very. " Atmospheric CO_2 is taken in by plants. During the process of photosynthesis the carbon is incorporated into carbohydrate building blocks that are used in plant functions and plant parts. Some refer to this process as carbon fixation. It has come into greater focus now with the increasing interest in climate change and consequences of increased levels of CO_2 in the atmosphere. As CO_2 is fixed, oxygen is released by the plant to the atmosphere as a by-product of photosynthesis. When plants die, their parts fall to the ground and ultimately are broken down by macro- and microorganisms whose respiration takes the carbon from the plant parts, converts it back into CO_2 and releases it to the atmosphere. Furthermore, some plants are eaten by animals, which digest the plant parts using them for energy, releasing CO_2 in the process of respiration, and also building body tissue from some of the plant parts. When animals die they also fall to the ground and are decomposed by macro- and microorganisms, resulting in CO_2 being released to the atmosphere (Figure 8-1).

The rate of decomposition of the bodies of dead plants and animals, and therefore the rate of release of CO_2 to the atmosphere, can vary greatly based on where the parts fall and the type of ecosystem in which they exist. If decomposition does not keep up with the accumulation of dead plant and animal material, this material builds up. In this situation carbon in the organic matter from dead plants and animals builds up in the ecosystem, or is "sequestered". Moreover, the organic matter falling to the ground may undergo partial decomposition but then enters the soil as organic matter and final decompositions occurs very slowly over many years. Therefore if organic matter builds up in an ecosystem faster than it breaks down, atmospheric CO_2 will decrease.

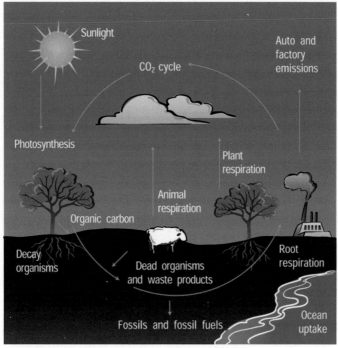

Figure 8-1   Schematic of the carbon cycle.

There is almost no data on the role vineyards play in carbon sequestration and greenhouse gas (GHG) production. Nitrous oxides (NOx) are also important GHGs. There is some indication that NOx produced in vineyards will play a bigger role in the consequences of GHG production because it is about 300 times more reactive in the atmosphere than is CO_2. Nevertheless, it is important to know the basics of the carbon cycle in your vineyard. For example, tillage of the vineyard floor exposes soil organic matter to microbes, greatly increasing its rate of decomposition and thereby the release of CO_2. The warmer the environment the faster the decomposition will occur. So if you are interested in enhancing the carbon sequestration in your vineyard, use a no-till vineyard floor management strategy. This will be discussed further in Chapter 11, Sustainable Soil Management.

WATER CYCLE

Water enters the landscape through rainfall and is stored in the soil profile, as surface water or as ground water in aquifers. Water cycles out of the landscape through runoff, evaporation, transpiration and deep percolation. These four components of the water cycle are strongly affected by the plants that cover the soil surface in natural and agricultural ecosystems. You can optimize your on-site water resources by reducing runoff, improving infiltration, and increasing soil water-holding capacity. Similarly, you can conserve water and protect water quality by minimizing off-site impacts, particularly the off-site movement of sediment (Reeves 2008). The Sustainable Water Management chapter will discuss some of the ways a grower can affect the water cycle on their farms.

ENERGY FLOW

Energy from the sun through plants is a one-way flow and therefore there is no actual Senergy cycle. Our natural living world runs on solar power and as an ecosystem process, energy flow shapes how ecosystems are structured and function. Our management decisions can affect how much energy is captured and put to use (Reeves 2008). Plants capture light energy, and through the process of photosynthesis, convert that light energy into stored chemical energy that ultimately is utilized by animals and us. All life depends on this and therefore so does every economy, every nation, and every civilization (Savory and Butterfield 1999). Understanding energy flow and learning how to enhance it without large inputs of fossil fuel products will improve the other three ecosystem processes. Richard Smart's book *Sun-*

light into Wine is an explicit recognition of the role sunlight plays in producing high quality wine (Smart and Robinson 1991). Canopy management is one way a grower can affect the energy flow in their vineyard.

COMMUNITY DYNAMICS

A community is a subset of the living organisms in an ecosystem. For example, in your vineyard there is a plant community and an animal community. The animal community can be broken down further into the soil microbial community, the insect community, the bird community, and so forth. Community dynamics can be thought of as how the communities in the ecosystem interact and change over time. These interactions are the most vital of the four ecosystem processes. The other three processes cannot function to our benefit unless plants of some form convert sunlight to useable energy for life and populate the environment.

Biologically diverse communities are very dynamic, in other words they are in a constant state of flux. Species composition, numbers, and age structure are changing constantly along with a variety of other factors within the community. Biological diversity, or biodiversity, is a measure of the number/variety of species of plants and animals in an ecosystem. A biologically diverse assemblage of plants and animals enhances the functioning, stability, and productivity of our environment (Boyce and Haney 1997, Di Giulio et al. 2001). A high biodiversity is complementary and essential to agriculture productivity and crop quality (Altieri 1999, Jackson and Jackson 2002, Imhoff and Carra 2003, Long and Pease 2005, Marshall et al. 2005, Blann 2006).

Succession is the term given to the process of changes a community and/or ecosystem experiences over time. Biodiversity changes during the process of succession and is therefore an important element to monitor within not only vineyards, but also adjacent natural areas on the farm. Relatively low biodiversity would indicate a stagnant successional process that could ultimately contribute to increased pest problems requiring more management and impact your bottom line.

The biological communities in an agricultural ecosystem provide benefits over and above the commercial crops they produce. These benefits are known as ecosystem services. They include removing carbon dioxide from the atmosphere, reducing greenhouse gases, the recycling of nutrients, and regulation of microclimate and local hydrological processes. In some cases they result in the suppression of

pest plants and animals through the production of pest natural enemies, and detox-
ification of noxious chemicals that enter the environment (Altieri 1999).

WE LIVE AND FARM IN A WATERSHED

Not only do we farm within an ecosystem, this ecosystem occurs within a watershed. A watershed is an area of land drained by a river or stream. Drained by a river
or stream means that when rain falls anywhere in that watershed the water will ultimately end up in the river or stream. The size of the watershed can vary depending
on one's definition. For example one can talk about the Mississippi River watershed, which encompasses a huge portion of the US, or break it down into smaller
ones such as the Missouri River watershed, and so forth. In California alone there
are 109 unique watersheds that have been identified by the CalFed Program (Cal-
Fed 2006; Figure 8-2). These are primarily major watersheds and do not include the
numerous smaller tributaries of these unique and valuable resources.

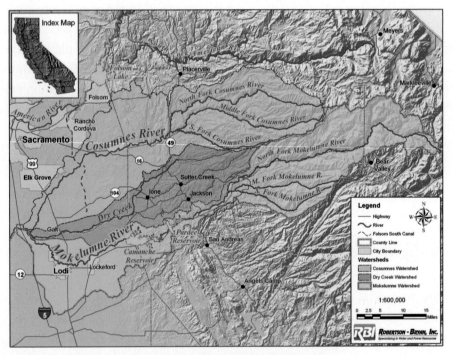

Figure 8-2   Map of the main watersheds in the Lodi region.
Courtesy of Lodi Winegrape Commission.

Some ecologists feel that a watershed is the minimum ecosystem unit (e.g. Odum 1971). I like to think of an ecosystem unit as a system defined by its plant populations, for example an oak woodland ecosystem, a grassland ecosystem, or a conifer forest ecosystem. The plant and animal species and how they interact in each one of these is very different. If one uses my definition, it is possible for a watershed to have several ecosystems within its boundaries. Some ecologists would call my ecosystems plant communities.

In any case, it is important for you to realize you live within a watershed and that activities in any given part of the watershed can have significant impacts on people and ecological communities in other parts of the watershed, particularly downstream. Because of the interrelationships of its various parts, a watershed is probably the best unit to use when doing long-term planning for a region. Agriculture is critical to the management of watersheds, from the participation of individual farmers in watershed stewardship efforts to the crops that are produced in watersheds. Healthy watersheds make for healthy working landscapes. "We all live in a watershed" is not just a catchy phrase repeated by watershed-based management advocates; it is the truth. Watershed stewardship groups are forming in more and more regions and if one exists in your area, I highly recommend you participate in it.

I would consider long-term watershed planning as the only way we are going to tackle critical issues such as land use, energy consumption, air quality, and water use to end up with the types of communities in which we all would like to live. However, most regional planning is still done on too small a scale, definitely not using the watershed approach. Political boundaries are usually what are used in regional planning, most of which have no relationship to watershed boundaries.

ECOSYSTEM MANAGEMENT ON THE FARM

A useful way to approach ecosystem management for your farm is to discuss it in terms of habitat management. Often what is labeled as a "habitat" is really a plant community. However, because many people from different professions, agencies, and organizations talk about habitat, it is important to use an accurate and consistent definition to facilitate effective communication among different professions, agencies, and organizations (Reeves 2008). Hall et al. (1997) define habitat as:

> ... the resources and conditions present in an area that produce
> occupancy—including survival and reproduction—by a given

> organism. Habitat is organism-specific; it relates the presence of a species, population, or individual (animal or plant) to an area's physical and biological characteristics. Habitat implies more than vegetation or vegetation structure; it is the sum of the specific resources that are needed by organisms. Wherever an organism is provided with resources that allow it to survive, that is habitat.

One of the aspects of this definition that I like is the connection between habitat and occupancy by organisms, both plant and animal. All of us desire to have certain occupants on our farm, whether they are our favorite flowering plants or wildlife. Having the right habitat is how to make this possible. Habitat that exists on the farm can be separated into that which occurs outside the vineyard as distinct from habitat within the vineyard.

VINEYARD AS HABITAT

The vines themselves provide habitat for certain animals. For example, in the Lodi region Mourning Doves commonly nest in the vines. The vine canopy also provides habitat for important insects and invertebrates such as spiders, which prey on any other invertebrate that moves and that they can overpower, including some of your important vineyard pests.

Figure 8-3   Bird nest in the canopy of a winegrape vine.

Vineyard habitat can be greatly enhanced by growing cover crops, which come in a huge variety (Ingels et al. 1998). Flowering cover crops provide nectar and pollen for parasites and predators of insect and mite pests (Brendt et al. 2006). They also provide organic material that enhances soil microbial communities. The importance of these communities in vineyard nutrient cycling and vine nutrition will be discussed in Chapter 11, Sustainable Soil Management.

Habitat on the edges of vineyards can be provided by planting grasses, trees and/or hedgerows. They provide a range of requirements for both vertebrates and invertebrates, such as food, shelter, and nesting sites. Bird boxes, raptor perches, and bat boxes can be put in and around vineyards to encourage populations of birds and bats. Owls, raptors and bats will prey on some of your important pests such as gophers, voles, and flying insects. During nesting season, songbirds require large numbers of insects to feed their young.

HABITAT OUTSIDE THE VINEYARD

The type of habitat that occurs outside the vineyard depends on a range of things, such as the climatic region, soil type, and attitude of the grower. In the Lodi region the range of habitats include individual trees, oak woodlands, riparian corridors, and vernal pools. Each one has its own unique characteristics, but all offer refuge, food, and nesting sites for vertebrates and invertebrates alike. Establishing a particular habitat or enhancing one that already exists will require practices unique to each situation.

ECOSYSTEM MANAGEMENT AND SENSITIVE SPECIES

There are approximately 360 plants and animals listed under the Federal and State Endangered Species Acts in California, a much higher number than in most other states (Reeves 2008). According to a 1993 study by the Association for Biodiversity Information and The Nature Conservancy, half of listed species have approximately 80% of their habitat on private lands. Because of listed species' dependence on private lands, private landowner participation in endangered species conservation is critical to successful species recovery and their eventual delisting.

The federal Endangered Species Act was created with a positive purpose in mind, to protect animals and plants that are so low in number, or live in such limited habitat, that any further deaths or reduction in habitat literally threatens their

Figure 8-4   Swainson's Hawk, an endangered species in Central California.

continued existence as a species. The US Fish and Wildlife Service enforces the act as it applies to terrestrial species. Violating the act can have serious consequences. Killing an endangered species can result in a fine of up to $50,000 or a prison sentence of up to one year (Reeves 2008). Special interest groups use the Endangered Species Act to accomplish a range of agendas in connection to the winegrape industry. For example, some use it as a means to fight vineyard expansion, while others use it to try to curtail specific vineyard operations, such as pesticide use. Unfortunately, many environmental regulations are reactive rather than proactive and as a result some unintended consequences can occur. In the case of the Endangered Species Act the three S's (shoot, shovel and shut up) evolved as a way for some landowners to deal with it. It comes into play when a grower discovers an endangered species on their property and, fearing the US Fish and Wildlife Service might find out and place severe restrictions on their farming practices, gets rid of the endangered species before anyone finds out it is there. Or if critical habitat or a food plant of an endangered species is found, the grower eliminates it so as not to attract the endangered species. While this approach to endangered species has become less common, it is still a problem.

It is important that growers realize that if an endangered species occurs on their farm it is because they are doing something right. It means they have created habitat that is good for these species and hopefully helps increase their numbers. More-over, several state and federal programs such as the Safe Harbor Agreement, provide mechanisms to protect landowners' interests, while providing incentives to manage lands in ways that benefit endangered species.

SAFE HARBOR AGREEMENTS

Fortunately, through the hard work and creative thinking of the Environmental Defense Fund and the US Fish and Wildlife Service, a program was developed to deal with the highly charged issues arising from the enforcement of the Endangered Species Act. The Safe Harbor Agreement has created a genuine win-win situation for the grower and the federal government. Basically, a Safe Harbor Agreement is a legal agreement between the landowner and the US Fish and Wildlife Service, assuring that no added Endangered Species Act restrictions will be imposed on the landowner as a result of the grower carrying out activities expected to benefit an endangered species. They effectively "freeze" a landowner's Endangered Species Act responsibilities at their current levels for a particular endangered species if he or she agrees to restore, enhance, or create habitat for that species. However, the agreement does not allow a landowner to deliberately harm endangered species on the property. Nevertheless, it does protect a grower from accidental "take" of an endangered species during normal farming operations or during restoration projects being done on the farm. Take is the official word used by US Fish and Wildlife Service that means killing an endangered species. The agreement may also contain language that allows a landowner to legally alter habitat of an endangered species. One of the key provisions of the Safe Harbor Agreement is that a survey of the property is done before the agreement takes effect to establish a baseline condition of the endangered species. When the agreement ends the baseline condition must be at least what it was at the start. The date on which the agreement ends is agreed upon by the landowner and US Fish and Wildlife Service before the signing. Since many endangered species are hard to count accurately, the baseline condition is often related to the amount of the species' habitat that is present at the start of the agreement.

Because of the highly charged nature of the topic of endangered species, some landowners may still be unconvinced as to how this could be beneficial to them. I will present an example to illustrate where it might come in handy.

More and more winegrape growers are exploring the idea of doing habitat restoration on their farms. There are several reasons for this increase in interest, and they differ among growers. Some are compelled to do it because they feel it makes them better land stewards. Others do it because the land development they are undertaking requires some kind of mitigation for habitat that is going to be lost during the development. Still others are interested because there are substantial funds available from programs like the California Bay-Delta Program, Wildlife Conservation Board, or the Federal Farm Bill for private landowners to do restoration work on their properties. In all cases, if the restoration work is being done in an area where endangered species habitat exists then the grower is liable for any endangered species habitat destruction or accidental killing that occurs as a result of the restoration work. I know of no grower who would go ahead with restoration work with this kind of liability hanging over their head.

The Safe Harbor Agreement was designed to overcome this obstacle to restoration work done in the presence of an endangered species or its habitat. The landowner and the US Fish and Wildlife Service create an agreement that allows the landowner to carry out the restoration work and protects them from the liability of habitat destruction and accidental killing of endangered species while the work is being done. An example in the Lodi area involves elderberry bushes, the food plant of the endangered valley elderberry longhorn beetle (Figure 8-5). A grower doing riparian restoration work may need to remove some elderberry bushes in the course of their work, which might also result in the accidental killing of some beetles. The Agreement allows the grower to do this work and not be liable for these actions. Of course the agreement also stipulates that the situation of the endangered species is at least the same, or better, when the agreement ends. In other words, at least as many or more elderberry bushes must be present in the project area when the agreement sunsets. Historically, the Environmental Defense Fund, which conceived of the Safe Harbor Agreement concept, has been the facilitator between the landowner and US Fish and Wildlife Service during the creation of these agreements.

The first Safe Harbor Agreement in the US was approved in 1995 to benefit the red-cockaded woodpecker in the sandhills of North Carolina. Since then 28 agreements have been developed in 15 other states, four of which are in California. One of the most recent Safe Harbor Agreements, and the only one to involve a vineyard, became effective September 10, 2003. At the time it covered Robert Mondavi Win-

Figure 8-5   Valley Elderberry Longhorn Beetle.

ery's new Cuesta Ridge Vineyard development on the Santa Margarita Ranch near San Luis Obispo, California. The agreement covered the reg-legged frog and two bird species

The evolution of the Safe Harbor Agreement provides some important lessons for the future. There is no doubt that winegrape growers will be confronted with more and more very serious regulatory issues that have the potential to limit their ability to farm as they have done in the past. History has demonstrated that these issues are not going to go away but will increasingly impact growers. Stepping back from the emotional aspects, issues such as endangered species are very important. For example, some species are ecological indicators and can be viewed as the "canary in the coal mine," indicative of a stressed environment. Given that environmental regulations can severely impact farming and also that the environment is being significantly impacted by human activities such as farming, creative solutions to these seemingly intractable problems must be developed. The Safe Harbor Agreement is one such solution.

Many landowners feel that all environmental groups are out to take away their livelihood. The reality is that some of these groups are beginning to recognize that their confrontational approaches to solving environmental issues on the farm, such as with lawsuits and getting stricter regulations passed by the legislature, are not accomplishing their goal of enhancing the environment. The unintended consequence of the Endangered Species Act discussed above is a good example. They are recognizing that one road to sustainable environmental management will come through cooperating with landowners to create working landscapes managed with sustainable practices. This kind of thinking led the Environmental Defense Fund to conceive of the Safe Harbor Agreement concept. It is working with growers, not against them.

CONSERVATION EASEMENTS

Conservation easements for protection of natural resources are legal agreements that allow landowners to donate or sell some "rights" on portions of their land to a public agency, land trust, or conservation organization. In exchange, the owner agrees to restrict development and farming in natural habitat, and assures the easement land remains protected in perpetuity. A 1996 study conducted by the National Wetlands Conservation Alliance indicated that the leading reasons landowners restored wetlands were to provide habitat for wildlife; to leave something to future generations; and to preserve natural beauty. Only 10% of landowners surveyed in the study restored wetlands solely for financial profit (Reeves 2008). This would also apply to other habitats besides wetlands. A conservation easement can provide financial benefits for the protection, enhancement, and restoration efforts for the natural environments on your property. The belief that natural resources such as wildlife, especially sensitive species, will reduce your land value is not true. Many easement programs include some sort of cash payment for a portion of the costs associated with habitat restoration and enhancement.

Agricultural conservation easements are for the explicit purpose of keeping farmland in production. They are similar to natural resource conservation easements, but, specifically protect farmland and maintain the practice of farming (Sokolow and Lemp 2002).

REFERENCES

Altieri, M. A. 1999. "The Ecological Role of Biodiversity in Agroecosystems." *Agriculture, Ecosystems, and Environment* 74:19–31.

Blann, K. 2006. "Habitat in Agricultural Landscapes: How Much is Enough? A State-of-the Science Literature Review." *Defenders of Wildlife*, Washington DC.

Boyce, M. S. and A. Haney, eds. 1997. *Ecosystem Management: Applications for Sustainable Forest and Wildlife Resources.* Yale University Press, New Haven CT.

Brendt L, S. Wratten, S. L. Scarratt. 2006. "The Influence of Floral Resource Subsidies on Parasitism rates of Leafrollers in New Zealand Vineyards." *Biol. Control* 37:50–55.

CalFed. 2006. Watershed program—final draft. http://www.calwater.ca.gov/content/Documents/library/ProgramPlans/2006/YR7_Watershed_ProgramPlan.pdf.

Di Giulio, M., P. J. Edwards, and E. Meister. 2001. "Enhancing Insect Diversity in Agricultural Grasslands: The Role of Management and Landscape Structure." *J. Appl. Ecology* 38:310–319.

Hall, L. S., P. R. Krausman, and M. L. Morrison. 1997. "The Habitat Concept and a Please for Standard Terminology." *Wildlife Society Bull.* 25(1):173–182.

Imhoff, D. and R. Carra. 2003. *Farming with the Wild: Enhancing Biodiversity on Farms and Ranches.* Sierra Club Books, San Francisco CA.

Ingels, C., R. L. Bugg, G. T. McGourty, and L. P. Christensen. 1998. *Cover Cropping in Vineyards. A Grower's Handbook.* University of California Div. Agric. Nat. Res. Publ. 3338. 162pp.

Jackson, D. L. and L. L. Jackson. eds. 2002. *The Farm As a Natural Habitat: Reconnecting Food Systems with Ecosystems.* Island Press, 1718 Connecticut Ave. Suite 300, Washington DC, 2009.

Long, R. and C. Pease. 2005. "Farmscaping with Native Perennial Grasses." *Grasslands* Spring 2005:6-7. Calif. Native Grasslands Assoc., Davis, CA.

Marshall, E. J. P., T. M. West, and D. Kleijn. 2005. "Impacts of an Agri-Environment Field Margin Prescription on the Flora and Fauna of Arable Farmland in Different Landscapes." *Agric. Ecosystems & Environ.* 113:36–44.

Odum, E. P. 1971. *Fundamentals of Ecology.* W. B. Saunders Co., Philadelphia. 574pp.

Reeves, K. 2008. "Ecosystem Management." *In* Ohmart C. P., C. P. Storm and S. Matthiasson. 2008. *Lodi Winegrower's Workbook.* 2nd Edition. Lodi-Woodbridge Winegrape Commission, Lodi CA. pp. 15–63.

Savory, A. and J. Butterfield. 1999. *Holistic Management: A New Framework for Decision Making.* 2nd Edition. Island Press, Covelo, CA.

Smart, R. and M. Robinson. 1991. *Sunlight into Wine: A Handbook for Winegrape Canopy Management.* Winetitles, Adelaide, Australia. 88pp.

Sokolow, A. D. and C. Lemp. 2002. "Agriculture Easement Programs…Saving Agriculture or Saving the Environment?" *Calif. Agric.* 56(1):9–14.

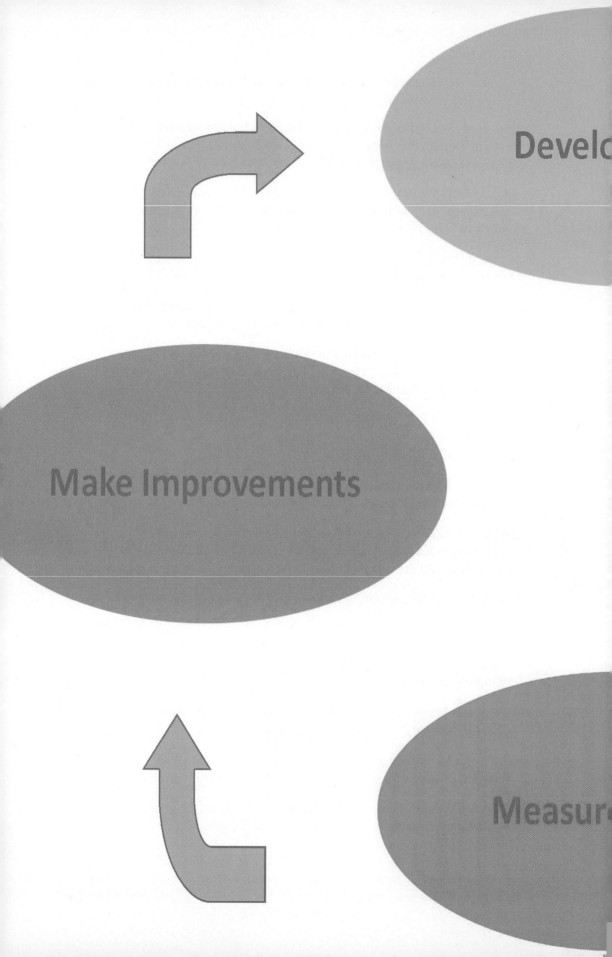

9

The Role of Self-assessment in Implementing Sustainable Winegrowing

In Chapter 1, I discussed the three challenges of implementing sustainable winegrowing: Defining it, implementing it in the vineyard on a day-in-day-out basis, and measuring the impacts of implementing these practices on the grapes, the vineyard and its surrounding environment. In 2000 the Lodi Winegrape Commission published the *Lodi Winegrower's Workbook: A Self-Assessment of Integrated Farming Practices* (Ohmart and Matthiasson 2000), a tool for winegrowers to use in meeting these three challenges (Figure 9-1). They recently published a 2nd edition, greatly expanding the workbook (Ohmart et al. 2008). The workbook not only had a significant effect on Lodi winegrape growers' approach to sustainable winegrowing but has also been a model for other wine industries in California, Washington State, New York, Michigan, Arkansas, Missouri, Oklahoma, and parts of Australia (e.g. Bernard et al. 2007; Brewer et al. 2009, Dlott et al. 2002; Johnson et al. 2009: www.vinewise.org; Wise et al. 2007; and *www.vinebalance.com*).

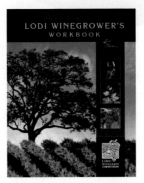

Figure 9-1   Cover of the *Lodi Winegrower's Workbook*, 2nd Edition. *Courtesy of Lodi Winegrape Commission.*

EVOLUTION OF THE SELF-ASSESSMENT WORKBOOK

Implementing sustainable winegrowing is to take the definition, such as the one discussed in Chapter 1, and translate it into farming practices used to grow wine-grapes. In other words, it is putting theory into practice. In 1998 Lodi Winegrape Commission growers came to the conclusion they needed a tool to help them expand their Integrated Pest Management program to encompass all aspects of growing winegrapes, i.e., transition to a sustainable winegrowing program. In my capacity as LWC's Sustainable Winegrowing Director, I reviewed other efforts to see how growers were meeting this challenge. Two programs stood out as ones having relevance. First, the Central Coast Vineyard Team (CCVT), working in the central coast winegrape region of California, had developed the Positive Point System (PPS) that a winegrape grower could use to assess the level of sustainable practices being used in their vineyards (Akerman et al. 1998). Second, Farm*A*Syst, based at the University of Wisconsin, had partnered with producers in the US, Canada, and Australia, to develop self-assessment workbooks for dairy, cotton, and other crops (http://www.uwex.edu/farmasyst/).

The Farm*A*Syst workbooks helped growers do several important things:

- Identify farming practices that were beneficial from an environmental perspective.
- Identify farming practices that were having a negative impact on the environment.
- Create action plans and a time table to address the practices causing environmental concern.
- Provide information in the workbook to help develop and carry out the action plans.

ENVIRONMENTAL MANAGEMENT SYSTEMS

Farm*A*Syst's approach to addressing environmental concerns on the farm is based on the Environmental Management Systems (EMS) or "Plan, Do, Act, Check" model, a standard process used to develop goals, implement them, measure success, and make improvements to ensure continuous improvement (Figure 9-2). EMS traces its roots back to the early 1970s and the United Nations Conference on Human Environment in Stockholm. One direct result of the conference was the formation of the World Commission on Environment and Development, which published *Our Common Future* in 1987 (World Comm. Environ. & Dev. 1987). This report identified worldwide environmental and social pressures and proposed

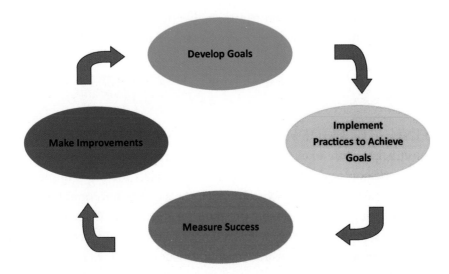

Figure 9-2 Schematic of the Plan, Do, Act, Check model for continuous improvement.

actions that would foster sustainable development. It was the first time the term "sustainable development" was used, and the report called for industries to develop effective "environmental management systems." EMS became the accepted name for an environmentally focused "Plan, Act, Do, Check" process.

Over the next several years, more and more world leaders publicly supported the report and in 1992, 172 nations participated in The United Nations Conference on Environment and Development in Rio de Janeiro, Brazil, which became known as the Earth Summit (http://www.un.org/geninfo/bp/enviro.html). Participants at the conference adopted a comprehensive set of guidelines for achieving a sustainable global environment. Furthermore, the international business community pledged support for the development of standardized management systems for environmental protection. It was during this period that the International Organization for Standardization (ISO) became involved. In 1993, ISO created a technical committee charged with developing a uniform international EMS standard, and in 1996, ISO released the "Standard for Environmental Management Systems," or ISO 14001 as it is most commonly called (http://www.iso14000-iso14001-environmental-management.com/).

A self-assessment workbook based on the EMS approach is well suited to a paradigm like sustainable winegrowing, where a continuum exists along which a grower can advance, becoming more sustainable over time. It is used to not only help a grower decide what practices to implement; it also shows the grower where they are, as well as where they should be headed on this continuum for each practice and for whole areas of farming. This will be discussed in greater detail below.

WRITING THE LODI WINEGROWER'S WORKBOOK

The project began with a key group of Lodi winegrape growers participating in a facilitated workshop to develop the principles and goals for the project. In other words they created a sustainable vision, much the same as one would do this as a first step in developing a sustainable program for one's farm. Once the principles and goals were established, a seventeen- member committee was formed, made up of Lodi winegrape growers, vineyard consultants who worked with Lodi winegrape growers, University of California (UC) Davis scientists as well as the regional UC Viticulture Farm Advisors, an NGO wildlife biologist, a US EPA Region IX staff member, the regional Natural Resources Conservation Service agronomist, the Director of Farm*A*Syst, and LWC staff.

Traditionally, Farm*A*Syst workbooks had been environmentally focused. However, the workbook committee felt it was important to also include issues related to quality winegrape growing and human resources. Therefore the workbook attempted to address all three "E's" in the definition of sustainable winegrowing, environmental soundness, economic viability, and social equity.

CREATING THE SELF-ASSESSMENT

The first step in creating the workbook was identifying all of the important issues/practice areas that are involved in growing winegrapes in the Lodi region, much the same as CCVT's PPS identified them for the California Central Coast winegrape region (Ackerman et al. 1998). Some examples are shoot positioning, plant tissue sampling, off-site water movement, monitoring the vineyard for insect and mite pests, soil erosion due to wind and water, and employee training. They were grouped into the following categories, which became workbook chapters: Viticulture, Soil Management, Water Management, Pest Management, Habitat, Human Resources, and Wine Quality. In the second edition, Viticulture was broken into two chapters, Vineyard Establishment and Viticulture; the Wine Quality chapter became Wine Quality and Customer Satisfaction; and a chapter on Shop and Yard

Management was added to deal with issues not directly related to the vineyard, such as recycling, hazardous materials management, and conservation in running the office and shop.

Once identified, each issue was put into a Farm*A*Syst worksheet template which, in my opinion, is what makes the Farm*A*Syst workbook model the most useful for implementing sustainable winegrowing. Rather than have each issue be addressed with a simple "Yes/No" question, the template requires that each issue be divided into four levels, called categories. One level, Category 1, lists the least desirable or least sustainable set of practices to address that issue. On the opposite end of the spectrum another level, Category 4, lists the most desirable or most sustainable set of practices to address the issue. The other two categories are between the two extremes, Categories 2 and 3 listing practices that are progressively more sustainable, in that order, between Categories 1 and 4.

Figure 9-3 provides an example issue worksheet from the Pest Management Chapter of the *Lodi Winegrower's Workbook* in order to illustrate several important points. The first is that each issue should be as specific as possible. In this case the issue is "managing omnivorous leafroller (OLR)."

Second, the categories go from 4 to 1, left to right across the page. This layout was done very deliberately. When filling out the workbook a grower is asked to read the worksheet for each issue and select the category that most closely describes what they do for that issue in their vineyard. People naturally read from left to right so a grower filling out the workbook will read Category 4 first and continue reading until they get to the category that matches what they do. In this way they read first what is most desirable for them to do from a sustainability perspective for that issue, i.e. Category 4. As one goes from right to left practices become more IPM intensive. If the issue worksheet was laid out Category 1 to 4, left to right, the grower would read Category 1, the least desirable practice, and likely will stop at the one that matches what they do and never read Category 4, the most desirable practice.

Third, the worksheets are constructed so that each farming issue, and the practices that go with it, are presented in a graduated form from least sustainable to most sustainable. The Farm*A*Syst workbook model puts each specific farming issue and each set of practices that address that issue onto a continuum. This does at least two things for the grower. First it helps them see where they are on the sustainability continuum for each vineyard issue. Second, it also helps them see where

6.9 Managing omnivorous leafroller (OLR)			
Category 4	**Category 3**	**Category 2**	**Category 1**
I do not have to treat for OLR because parasites keep the population below the economic threshold *Or* If control is necessary due to problems in the previous season, I use pheromone confusion for control.	I check 10 flower clusters on 20 vines at bloom time for treatment decision-making *And* I obtain biofix for OLR using a pheromone trap, and degree-days are tracked using weather station data *And* Treatment is timed for 700-900 degree-days from biofix (the most susceptible stage) *And* I use LWWC's PEAS model in selecting the pesticide to use for OLR taking into account the environmental impact units (EIUs) and efficacy.	My treatment for OLR is based on the time of year or stage of grapevine development (e.g. bloom) *And* I use LWWC's PEAS model in selecting the pesticide to use for OLR taking into account the environmental impact units (EIUs) and efficacy.	My treatment for OLR is based on the time of year of stage of grapevine development (e.g. bloom) *And* Environmental impact is not considered when I choose a pesticide to use for OLR.

Figure 9-3   Workbook worksheet for Issue No. 6.9 from the Pest Management Chapter in the *Lodi Winegrower's Workbook,* 2nd Edition (Ohmart et. al. 2008).

they should be going to become more sustainable—in other words, what practices to implement to become more sustainable on a particular issue. It is a road map of practices on how to get to Category 4 for each issue. Furthermore, most of the pest management issues can also be viewed through an Integrated Pest Management (IPM) lens. Going from right to left on the worksheet one goes from no IPM in category 1 to high-level IPM in category 4.

One of the things I like about this workbook model, in particular, is that it helps the grower deal with the all too human habit of denial. Because of the complexity of winegrape growing there is always something more that can be done to move along the sustainability continuum. Denial allows us to avoid the issues that we know we should tackle but do not. The workbook helps identify some of those issues in a very constructive way.

The final step in completing the self-assessment portion of the workbook was adding educational information about specific issues and practices. The workbook was not intended to be a textbook on winegrape growing, exhaustively covering all topics. However, the committee felt that certain facts should be in the workbook that would be beneficial to the reader, either as a stimulus for them to find out more about the topic in resources referenced in the workbook or useful for creating and carrying out action plans described below.

How To Use The Workbook

On each worksheet in the workbook, a grower picks the category that best describes the practices they use for that workbook issue. For example, if a grower treats for OLR based on time of year or stage of the grapevine development and uses the Lodi Wine-grape Commission's Pesticide Environmental Assessment System (PEAS) to select the pesticide to use for the treatment, then they are a category 2 (Figure 9-3). There is a summary evaluation sheet for each workbook chapter listing all the worksheet issues in the chapter and a series of boxes corresponding to each of the 4 categories. Figure 9-4 presents the first portion of the summary evaluation sheet for the Pest Management Chapter. In the example described above, the grower would check the box for category 2 for Issue No. 1. They then continue reading all the worksheets in the workbook, checking the appropriate boxes for each one. There is a Not Applicable (N/A) category available, because some workbook issues are not applicable to all nsvineyards.

Action Plans

The next task addressed by the workbook is to provide guidance in developing action plans for the areas of vineyard management identified by the self-assessment process as needing improvement. Since Category 1 is the least sustainable set of practices for a particular issue, any worksheets scored a 1 indicate areas of vineyard management

Issue	Pg No.	4	3	2	1
6.6 Economic threshold for leafhoppers	210	√			
6.7 Economic threshold for Willamette mites	211				√
6.8 Economic threshold for Pacific mites	211	√			
6.9 Managing ominivorous leafroller	214			√	
6.10 Managing grape leaffolder	217	√			
6.11 Mealybug management	218				√

Figure 9-4   Example Summary Evaluation Sheet for a portion of the Pest Management Chapter in the *Lodi Winegrower's Workbook*, 2nd Edition (Ohmart et al. 2008).

that should probably be addressed first. The summary evaluation sheets make it very easy to scan the complete list of issues, picking out ones that need the most attention.

The workbook contains blank templates for creating action plans, illustrated in Figure 9-5. It provides space to list the section of the workbook in which the issue occurs, the page where it is located, the area of concern, the plan of action to improve on the issue, and a time table for carrying out the action plan. The action plan is one of the most important components of the workbook. There is not much point in doing a self-assessment if one is not prepared to make improvements in vineyard practices based on the results. Referring back to being in denial about certain issues on the farm that should be addressed, if one is not going to make improvements in areas identified by the self assessment it is probably better to remain in denial, because at least one will not feel guilty about one's inaction.

MEASURING IMPLEMENTATION

If one goes to the effort of implementing sustainable winegrowing, then it is important to be able to measure its level of adoption. The four-category worksheet of the Farm*A*Syst workbook model is set up to do just that. It can be done for an individual vineyard and grower as well as for a group of growers in a region or larger geographic area.

An individual grower can use their evaluation sheets as a summary of their assessment. If they carry out one or more action plans they can assess their vine-

Action Plan				
Workbook Section	Issue Number	Issue and area of concern	Plan of action	Timetable for action
Pest Mgt: Insect and mite management	6.9	Managing OLRs: My treatment is based on time of year and I do not consider environmental impact when I select the pesticide to use	Monitor flower clusters for OLR, time my treatments, if necessary, using a biofix by using pheremone traps to catch adult moths and I will use the PEAS model in making pesticide selection	Next growing season

Figure 9-5   An example Action Plan from the Pest Management Chapter from the *Lodi Winegrower's Workbook*, 2nd Edition (Ohmart et al. 2008).

yard practices again after one or two seasons and track their improvements over time looking at the changes in their assessment "scores."

A group of growers in a region can pool their evaluations into a common database which can then be summarized. For example, for Lodi's workbook program I created a Microsoft Access database for capturing and summarizing vineyard workbook self-assessments. Growers willing to share their vineyard evaluations allowed them to be put into this database. In the first two years of the workbook program (2000–2001) 200 vineyard evaluations were entered into the database.

Assessment data can be summarized in several ways. One is to calculate an average "score" for each workbook issue. Figure 9-6 contains a bar graph for the issues in the first half of the Pest Management chapter with the height of the bar being the average category for that issue. For example, the average for Issue 1 "Monitoring for insect and mite pests" is 3. Category 3 on the worksheet for Issue 1 informs us that the average grower is monitoring their vineyards at least once a week for insect and mite pests, but not keeping a written record. By calculating the average score for each workbook issue, one can benchmark in a very detailed way where the region's winegrape growers are on the sustainable winegrowing continuum. One can readily see where growers are doing well and also where improvements are needed. One way I used these data summaries was to determine what farming issues require more attention in LWC grower outreach meetings. Furthermore,

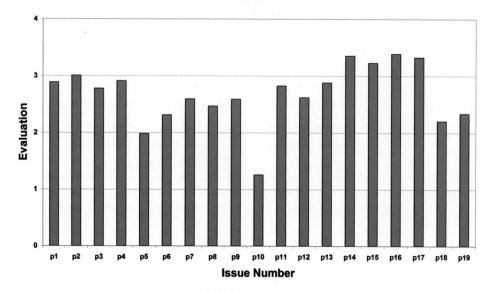

Average Evaluation from 200 Vineyards for Pest Management Chapter

Figure 9-6   Bar graph of the acreage scores from 200 vineyards for the first 19 Issues in the Pest Management Chapter of the *Lodi Winegrower's Workbook* (Ohmart and Matthiasson 2000).

with average scores for each issue in a region, an individual grower can then see how their score matches up to them.

Another way to summarize the data is to look at the proportion of 4's, 3's, 2's and 1's which were recorded for each issue. Figure 9-7 contains a stacked bar graph that displays this data for half the Pest Management chapter issues. In this manner one can see the portion of the growers doing specific practices. For example for Issue 1 "Monitoring for insect and mite pests," 20% of the growers are in Category 4, which means they monitor their vineyards at least once a week and keep a written record of what they see. The color coding of the bars makes it easy to get an overall sense of the level of implementation for a workbook chapter. For example, seeing a high proportion of blues (4's) and greens (3's) for most issues means that there is a high level of sustainable practices, whereas seeing a high proportion of reds and yellows indicates a low level of sustainable practices.

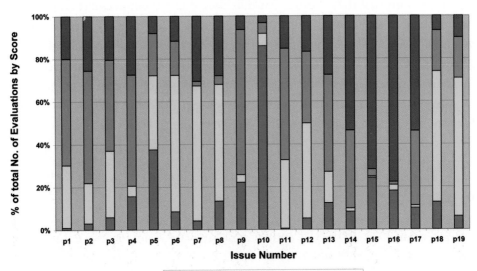

Figure 9-7   Bar graph showing the percentage of vineyards with scores of 1, 2, 3 or 4 for the first 19 Issues in the Pest Management chapter of the *Lodi Winegrower's Workbook* (Ohmart and Matthiasson 2000).

CONCLUSIONS

The paradigm of sustainable agriculture and sustainable winegrowing continues to evolve. As discussed in Chapter 1, no one will ever create a perfectly sustainable vineyard because there will always be something that can be done to make it more sustainable. I see this as a source of frustration for a winegrape grower, and maybe an impediment, because it is human nature to want to be able to arrive at the prize, so to speak. However, I think sustainable winegrowing has to be looked at as a process of continual education and improvement in practices. It is a journey, not a destination. Furthermore, each grower is going to be at a different level of implementation of sustainable winegrowing and may move along the continuum at a different rate.

A Farm*A*Syst self-assessment workbook is well suited to dealing with the situation described above. First, because it encompasses the complete range of practices for each farming issue, from less to more sustainable, it is suitable for most growers. Second, for every farming issue, it is a road map of practices showing a grower exactly where they are at in terms of a level of sustainability and what they can do to improve if they choose. Third, it encourages them to create and carry out action plans to make improvements. And finally, it provides a form of measurement that helps a grower track themselves through time, either individually or as a group. However, it is important to point out here that in the future, metrics around sustainable winegrowing will need to move past simply tracking practices and include performance measures, such as energy expended and gallons of water used per ton of grapes produced, as well as balancing multiple factors along with farmgate income.

REFERENCES

Ackerman, D. et al. 1998. "Central Coast Vineyard Team Positive Point System." *Practical Winery and Vineyard* 29(1):12–24.

American Society of Agronomy. 1989. "Decisions Reached on Sustainable Agriculture." *Agron News* January. p 15.

Brewer, M. et al. 2009. *Grape*A*Syst: Michigan Grape Grower Sustainability Assessment and Risk Reduction Tool.* Mich. St. Univ. Ext. East Lansing, MI. 75pp.

Dlott, J., C. P. Ohmart, J. Garn, K. Birdseye, K. Ross, eds. 2002. *The Code of Sustainable Winegrowing Practices Workbook.* Wine Institute & Calif. Assoc. Winegrape Growers. 477pp.

Johnson, D. T., R. K. Striegler, R. A. Allen, R. J. Smeda, E. A. Bergmeier, J. L. Harris, and J. Cotta. 2009. *Ozark Mountain Vineyard Sustainability Assessment Workbook: A Self-Assessment of Management Practices.* University of Missouri Extension. 77pp.

Ohmart, C. P., Matthiasson S. 2000. *Lodi Winegrower's Workbook: A Self-Assessment of Integrated Farming Practices.* Lodi-Woodbridge Winegrape Commission, Lodi CA. 145pp.

Ohmart, C. P., C. P. Storm, and S. K. Matthiasson, eds. 2008. *Lodi Winegrower's Workbook.* Lodi Winegrape Commission, Lodi CA. 350pp.

Wise, A., Martinson, T., Hawk, J., Weigle, T., Tarleton, L. 2007. *New York Guide to Sustainable Viticulture Practices: Grower Self-Assessment Workbook.* Cornell University Cooperative Extension. *www.vinebalance.com.*

World Commission on Environment and Development. 1987. *Our Common Future.* Oxford University Press. 398pp.

10

Data-Driven Sustainable Winegrowing: A Case for Good Record Keeping

A great deal of research is going on in the field of sustainable winegrowing. In visiting vineyards, talking to winegrape growers, and reading various trade magazines, it is clear we have a long way to go in terms of developing science-based (i.e., data-driven) sustainable winegrowing practices that are widely adopted in the vineyard. There are many reasons why this is the case, but some of the most important factors are time, money, and a shortage of each.

In reflecting on my talks with growers and wine makers around the country about various approaches to winegrape growing and wine making, it struck me that great wines have been made using all types of farming methods, whether organic, sustainable, biodynamic or "conventional." Nevertheless, this does not stop some practitioners in each group from talking as if they have found the secret to fine wine making. Conversely, poor quality wines are produced using each of these different farming methods, too. So what does this mean? For one thing, no one approach to farming has been shown to provide *the* answer to growing quality winegrapes and making quality wine. Furthermore, since one can arrive at a high quality wine from many different directions, it also means that there is a lot more to learn about growing quality winegrapes and making quality wine. In other words there is a lot of variability in the processes of winegrape growing and wine making that we cannot explain yet. There are likely many formulae for producing great wines.

Rather than first focusing on what we do not know about growing quality winegrapes and the variability that we cannot explain, let's look for common threads that run through the various farming approaches (i.e., conventional, sustainable, organic, biodynamic) when it comes to successfully producing high quality wines. One common thread is vine balance. Even though we might not all agree on the

Figure 10-1   A well balanced Cabernet Sauvignon canopy midway through veraison, with a suitable amount of sun-exposed fruit, uniform lignifications, and an open canopy.

same crop to pruning ratio, number of vines per acre, number of buds per cordon, and so forth, most viticulturists agree that quality winegrapes are produced from a balanced vine, one with the right amount of fruit produced per unit of canopy for a particular site. How does one achieve good vine balance? By recognizing that vine balance is affected by many things, such as soil type, soil quality, vine density, yield per vine, root stock, clone, trellis system, irrigation practices, shoot positioning, buds per spur, spurs per foot of cordon. Many of these factors are addressed during site selection, vineyard design and development, and early vine training. Once the vineyard is established, then close attention must be paid to the vines and the canopy manipulated if proper balance has not been achieved. Fine wines are produced when growers pay very close attention to their vines, regardless of whether the farming techniques are conventional, organic, biodynamic, or sustainable, giving credence to the expression "the best thing you can put on your vineyard is your own shadow."

Now let's focus for a moment on what we do not know regarding the growing of quality winegrapes and making fine wines and what we can do about it. One thing winegrape growers can do is to become better record keepers and to be more quantitative in their measurements. Winegrape growers have a long way to go when it comes to keeping good records and using them to make future management decisions. By record keeping I do not mean pesticide use, payroll, equipment, and other basic financial account records. Hopefully, most growers are pretty good at keeping track of these aspects of farming. I am referring to the "accounting" of the specific cultural practices like shoot thinning, shoot positioning, petiole samples, fertilizing, pest monitoring, neutron probe readings, weather data, and pressure bomb readings. Why is good quantitative record keeping so important? Because good records tell you very specifically what you have done in the past and, with careful analyses, they can be used to direct future practices. For example, vine balance is affected by many variables, and it is only through careful measurement of these variables and analyses of the data that patterns will emerge as to which ones have the most influence on the production of quality winegrapes. It is only through good record keeping that we can develop cost/benefit analyses of specific farming practices.

There are several reasons why growers have up until now been deficient in good record keeping. First, many farming practices are hard to quantify in an affordable and meaningful way. Pest monitoring is a great example. Most growers, and many pest management consultants are not trained to develop meaningful quantitative pest monitoring programs and many that have been developed by University scientists are too time consuming to be useful on the farm. Second, the value of good record keeping is not readily apparent in many cases, particularly to a small grower. It becomes more apparent when comparing one grower's practices with another's, which many growers are reluctant to do, or when considering a grower that farms multiple vineyards. Keeping good records of practices done on different varieties, in different aged vineyards, on different soil types, for example, and then comparing the resulting winegrape quality will most likely reveal patterns from which one can learn.

Another important reason why few growers keep good records of vineyard management practices is the lack of commercially available, affordable, easy to use computer database software to handle this kind of data and an easy and inexpensive way to collect the data in the field and get it into the database. The technology that can be used to develop the tools needed for good record keeping has been

Figure 10-2   Monitoring for pests with a hand lens.

around for a while, but the demand for it has not been sufficient to stimulate private companies into developing useful, commercially available products.

Some privately developed systems exist. For example, I developed a relational database in 1995 that was used for the ten years in Lodi Winegrape Commission's grower program to track all the activities for 70 vineyards farmed by 45 different growers. It is set up to easily enter, edit, and summarize all types of vineyard management data with a few mouse clicks and keystrokes. A personal digital assistant (PDA) was programmed and used for easy data entry in the field and subsequent entry into the computer database.

Commercially available software can be used to record pesticide use data and generate pesticide use reports. That is because in California growers are required by law to report to their county Agricultural Commissioner's office on a monthly basis records of all the pesticides they use. This created a commercial demand for this kind of software to be developed. Furthermore, there are some companies providing software, hardware, and services for doing GIS/GPS data collection and analyses. However, it is pretty pricey and its utility is limited so far. For example, data collection devices are commercially available that a grower or pest management consultant can use to record pest counts at specific GPS points in the vineyard. Since few growers and consultants have collected quantitative pest data in the past, like leafhopper nymphs per leaf, for example, having a pest count at a few specific locations in the vineyard is not that useful in pest management decision-making. That is because we don't know what these numbers mean in terms of the entire vineyard and also what they mean in terms of impact on the crop. However, the more we use these tools and record the data, the better we will become at interpreting it.

For winegrape growers to become better record keepers they need to be convinced that these records are valuable because they will help them produce more and higher quality winegrapes, save money, earn more money, or all of the above. The collection and systematic analyses of these records is the only way I see us getting away from grower testimonials being the major method for validating sustainable winegrape growing practices. I think we are about to see this happen. Just in the last five years several companies have developed and are marketing vineyard database software. Affordability is one of the big road blocks for many growers. However, if a product is viewed as valuable by a grower, then the amount of money they are willing to spend on it will go up. Getting growers to value data is a key factor in their willingness to invest in collecting it. In some ways it is a Catch 22 situation. Growers will not collect data until they value it but they will not learn to value it until they collect it and use it.

There is an old saying in the manufacturing industry: if you can't measure it you can't manage it. In the past, people did not become farmers because they wanted to be great record keepers. However, the global marketplace is putting a huge amount of pressure on growers to reduce their cost of production and track their vineyard inputs such as water and energy, and track their greenhouse gas production. Good measurements, efficient record keeping and good data analyses are important ways to ensure production is optimized. Moreover, a large part of sustainable winegrowing is attention to detail. This will become clear in the following chapters.

Part III

Practicing Sustainable Winegrowing: From the Vineyard

11

Sustainable Soil Management

Soil is the foundation of the vineyard. It not only provides the substrate in which the vines grow but also is the receptacle from which the vines get their water and nutrients. About the only thing vines do not get from the soil is the carbon dioxide that they need for photosynthesis. Sustainable management of soil is not only environmentally sound but is a good business practice. Gaining a greater appreciation for the soil resource in your vineyard is critical to make informed decisions on appropriate pre-plant operations and ongoing management. Knowing your soil resource gives you greater control over yield and quality and is especially important in determining the long-term sustainability of the vineyard. Furthermore, soil is also thought to be a major contributor to wine character, i.e., wine terroir.

Terroir is a concept that is much debated and is a term used to describe the characters a geography bestows to the wine. Geography encompasses the climate, soil, and all the other physical characters of the vineyards site.

As you will see from the following paragraphs, there is nothing magical about managing soil sustainably. As with all vineyard practices, sustainable soil management is understanding the soil and the important processes that occur in it, having a clear vision of what you want from your vineyard, and monitoring important indicators that will help in management decision-making. Books have been devoted to the study of soil and how it interacts with vines growing in it (e.g., White 2009). I will not attempt to present an exhaustive recreation of these texts here but simply highlight some of the important things to keep in mind when thinking about managing your vineyard soil.

Figure 11-1　Soil pit dug in between the vine rows.
Courtesy of Lodi Winegrape Commission.

SOIL BASICS

Several soil characteristics are important for maintaining long-term productive soils and grape quality and vine growth (Horwath et al. 2008):

1. Sufficient organic matter content to support microbial diversity (amounts will be texture dependent).

2. Soil depth of greater than 60 cm (depending on soil texture).

3. Soil should have good texture or organic matter level to support active root growth and water infiltration.

4. Maintain available water (about 2 inches, 0.6 mPa) in the root zone.

5. Have good aeration following drainage from irrigation or rain (minimum 15% to 25% air in soil pores).

6. Have a pH of 6 to 7.5 in the root zone and above 7 in the surface soil to maintain nutrient availability and minimize root disease for California soils. In vine-

yards in the Northeastern US it is rare to encounter a soil with a pH of greater than 6.5 and a pH of 5 is acceptable in some cases (Mark Chien and Tim Martinson, pers. comm.).

7. Enhanced soil biology by means of amendments such as cover crops and composts.

8. Acceptable salinity levels.

Soil is the most complex ecosystem on earth. One of the greatest benefits of soil organic matter is its effect on soil biodiversity. Generally, increasing amounts of soil organic matter will lead to increasing biomass and diversity of soil organisms. The great amount and diversity of soil life influences ecosystem processes, such as the carbon cycle and nutrient availability, which control the productivity of the vineyard. Moreover, a diverse soil biological community can often reduce the incidence of soil-borne diseases (Horwath et al. 2008).

Many things influence the biology of the soil, such as the presence of cover crops, tillage, rainfall, irrigation, compaction, and application of pesticides and fertilizers. The various groups of organisms in the soil perform a variety of activities including litter breakdown and decomposition (nutrient cycling) and killing other soil organisms through predation, parasitism or disease. The actions of these organisms, together with their by-products, excreta, and dead bodies, promote the dynamics of soil organic matter formation and turnover, mediate the availability of nutrients, and change soil structure. The soil food web is greatly influenced by the quantity and quality of plant residues and soil amendments entering the soil. That is why activities such as cover cropping and adding compost are important considerations that influence the outcome of soil ecological processes that impact vineyard productivity (Horwath et. al. 2008).

At least a third of the grapevine lives underground in the form of roots. Roots provide the pathway to move soil resources such as water and nutrients into the grapevine. Leaves feed the vines sugar produced via photosynthesis, but the roots provide the grapes all the essential components such as nutrients and water that provide the building blocks for photosynthesis and influence quality and some of the factors affecting wine character. The soil provides three vital things in order for roots to be effective: water, nutrients, and air. These three elements are best provided by a soil with good depth and structure—i.e., a soil in which the particles are bound together into small clumps, called aggregates, of varying size. Soil aggregation is a measure of soil structure. Soil organic matter is important in maintaining soil structure because byproducts from microbial breakdown of organic matter glues soil minerals together to form aggregates. Spaces between large aggregates permit rapid drainage and easy root growth, and spaces between small aggregates trap water for use between irrigation and rain events. One of the more important aspects controlling aggregate stability is the amount of microbial activity and soil organic matter. Stable aggregates occur in varying sizes and are created by the cementing action of microbes and their byproducts and soil organic matter. The assemblage of soil aggregates creates habitat to promote faunal and microbial diversity, an important index of soil quality. Management of soil structure is done primarily through additions of organic amendments such as compost, growing cover crops, and minimizing tillage (Horwath et al. 2008, Ohmart et al. 2000).

SOIL ORGANIC MATTER

Maintenance or enhancement of soil organic matter requires a consistent management strategy and one of the best is through the growing of cover crops. Cover crops add vital components necessary to feed microbial activity, one of the most important factors regulating the amount and persistence of soil organic matter. There are many different kinds of cover crops (Ingels et al. 1998) and they can be grouped according to what they contribute to the soil. For example, legume-based cover crops add more nitrogen than organic matter while cereal species and grasses primarily add organic matter. In fact cereal and grass species can tie up nitrogen and mixtures containing them can be used to control the amount of nitrogen supplied by cover crops to avoid problems in soils with excess nitrogen. Cover crop strategies have been developed to make nutrients readily available and maintain soil organic matter (Ingels et al. 1998). Legumes are often planted with grasses, because of their complementary functions, e.g., nitrogen fixation.

The response time of soil properties as a result of the growing of cover crops varies. For example, in the Mediterranean climate of California the sustainable nutrient cycling benefit is seen on average within three years as a result of the accumulating soil organic matter through the use of cover crops and soil amendments. The results may be much sooner in the moister temperate conditions in vineyards in the Northeastern US (Mark Chien, pers. comm.). Changes in soil tilth and water infiltration are often evident after only one or two years (Horwath et al. 2008).

Growing cover crops in vineyards has other benefits, including (Horwath et al. 2008):

- Reduced soil crusting
- Reduced winter runoff and soil erosion
- Better equipment access to vineyards after rain
- Increased humidity and cooler soil temperatures
- Suppression of weeds
- Promotion of beneficial insects

Depending on the amount and frequency of annual rainfall it is important to remember that cover cropped soil may require additional irrigation to replace soil water used by the cover crop. They also may compete with the vines for nutrients. Grass cover crops may require some added nitrogen whereas legumes may require phosphorus and sulfur. However, depending on the vigor of the vineyard and wine quality goals competition for water and nutrients by the cover crop may be used to your advantage. One final comment about managing a permanent cover crop is that after several seasons the soil may require some aeration. There are some implements that will aerate the soil while causing minimum disturbance to the cover crop and soil surface.

Composts are another good way to manage soil organic matter, particularly to enhance soil structure and water holding capacity. Once established, compost additions should be scaled back to avoid any potential problems with salinity issues. Composts derived from lawn wastes are normally low in salts, while composts made with food waste or manure can have high salt concentrations. Though vines are not particularly sensitive to salts relative to other crops, their growth will be impacted with salinity levels in the range of 2.5 to 4 ds/m. Problems with soil salinization can occur from repeated or over application of composts. It is best to use a combination of cover crops and composts, especially in soils of low organic matter content. In higher quality soils, cover crops are often sufficient to build soil organic matter and aggregates (Horwath et al. 2008).

Composts have a wide range of nutrient availability depending on their quality. Generally well-aged composts are low in nitrogen availability but can be a source of phosphorus and potassium. Nitrogen availability in composts depends on the age of the compost and the starting materials from which it was derived. Younger (less

Figure 11-2   Mowing a permanent resident vegetation cover crop during the early spring in a very old Zinfandel vineyard.

aged) composts have more available N. High nitrogen feedstock, such as food waste or manure added to the compost can also increase nitrogen availability. When purchasing compost you should scrutinize its consistency and ask for data on its nutrient value. If you do your own composting, determine the amount of nutrients it is contributing to your vineyard so your nutrient budgeting is as accurate as possible and you do not over or under apply nutrients (Horwath et al. 2008).

Composts can be especially important when establishing vineyards that have soils low in organic matter. Compost additions can also be used for maintaining good surface soil properties and as a mulch to prevent water evaporation. Other benefits of soil organic matter include:

- The buildup of beneficial soil organisms like earthworms through providing habitat and food
- Slow release of nutrients, particularly nitrogen and phosphorus
- Increased cations in the soil surface (NH_4^+, Ca^{2+}, K^+, Mg^{2+})
- Competition among organisms to suppress some root pathogens
- Chelated nutrients such as iron and zinc to make more available to the vine
- Buffers soil pH requiring fewer amendments

Most vineyard soils have organic matter contents of 2 to 4%. Depending on soil texture, soils with less than 2% organic matter should be managed to increase it with cover crops and compost. Soils with sandy texture are inherently low in soil organic matter with concentrations ranging from below 0.3% to less than 1%. Soils low in organic matter have a poor capacity to buffer soil pH. In a few cases, excess soil organic matter can reduce the availability of Zn and copper by incorporating them into its structure. In California it is difficult to increase the percentage of soil organic matter because year-round soil biological activity tends to reduce it. The most effective way to increase soil organic matter is to use cover crops and compost as described above. However, in some situations, such as the vineyards in the Northeastern US, too much compost addition can result in excess vigor problems due to the presence of excess nutrients from the compost (Mark Chien, pers. comm.).

SOIL STRUCTURE

Despite what some sales representatives of soil additives may tell you, no one product or technique has been developed that can mimic or accelerate the slow process of soil organic matter maintenance and soil aggregate formation. It is important to know that aggregates are susceptible to damage from disturbance or lack of organic matter input resulting in poor soil structure. They can be damaged or destroyed in a short period of time through improper or excessive tillage and compaction. Loss of soil structure will result in roots being deprived of water, nutrients, and air due to a loss of pore spaces in the soil and its reduced ability to hold nutrients. Vines with roots growing in a soil with good structure are able to cope better with soil pests and diseases and other stresses.

Adding organic matter through the growing of cover crops and compost application can overcome poor wetting patterns by improving water infiltration and making it more uniform. A more uniform distribution of irrigation water from drip systems can be achieved by building soil structure with cover crops and organic amendments such as compost (Fig. 11-3). The more uniform pattern results in better, more uniform soil moisture and assists the root systems in accessing available nutrients, permitting the use of less fertilizer and/or its efficient uptake. Efficient uptake of nutrients will help grow a more uniform grape crop, ensuring higher wine quality (Horwath et al. 2008).

The use of microbial inoculants and humic-based products has not been demonstrated to be effective in managing soil organic matter. Moreover, the use of these products in promoting vine growth and vigor is not established in the scientific literature. There are some exceptions to this rule. Sandy soils low in organic matter may benefit from having the rootstock inoculated with effective mycorrhizal spores. Mycorrhizae are fungi present in all soils and benefit roots by forming symbiotic relationships with them that promote nutrient and water uptake. They can also protect roots from pathogens. In return, mycorrhizae take sugars from the vine for their own maintenance. Inoculation of soil with mycorrhizae after the vineyard is established is generally less or not at all effective. If you feel compelled to use such products, carefully assess their potential effectiveness as a soil management practice. The literature on this subject is not well developed and it is advisable to get information on these products from more than one source (i.e., University of California Cooperative Extension, certified Pest Control Advisors, vineyard consulting businesses, etc.) (Horwath et al. 2008).

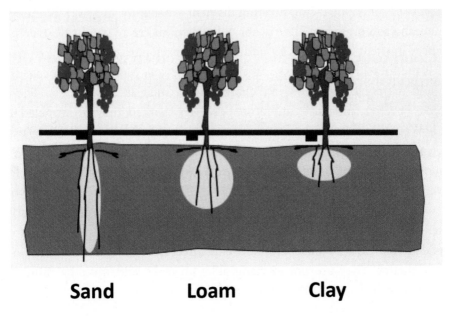

Sand   **Loam**   **Clay**

Figure 11-3 Irrigation wetting patterns under different soil textures.
Courtesy of Lodi Winegrape Commission.

Frequent tillage may lead to loss of soil organic matter through disruption of soil aggregates and stimulation of soil organic matter breakdown under conditions of low organic matter inputs. Under conditions of high organic matter input, tillage can be beneficial to create soil contact with microorganisms to increase organic matter breakdown. Once the desired soil organic matter level is achieved, reducing tillage will increase the organic matter content at the soil surface. This is desirable to enhance water infiltration and reduce dust emissions.

SOIL pH

One of the most important benefits of maintaining proper soil organic matter levels is its effect on soil pH, one of the more useful indicators of soil quality. Proper pH ensures availability of macronutrients (N, P, K, S, Ca, Mg, K) and micronutrients (Fe, Mn, Cu, Zn, B, Mo). All nutrients required by the grapevine are available at pH from 6 to 7.5. Under acid soil conditions (pH<5), aluminum toxicity may impact vine vigor by reducing root growth. Acid soils can result in manganese toxicity, typified by leaf roll and marginal necrosis at the end of the season. Zinc and iron deficiencies can occur in alkaline soils (pH > 7.5) (Horwath et al. 2008).

Soil pH is only one factor affecting nutrient availability, however. Building and maintaining soil organic matter is not only useful to create good soil structure; it enhances nutrient availability by adding cation exchange capacity (CEC) to soil. CEC is the measure of the electrical charge of the soil. More charge means that it can hold more cations (positively charged ions, such as Mg, Ca, K). The soil's charge is negative, and since cations are positively charged, they are attracted to the soil. Soil organic matter content and its mineral properties, specifically clay content, control the CEC of soil. The CEC is therefore related to soil texture. The amount of CEC in soil organic matter is extremely high compared to what is found on minerals. Soil organic matter content is small; typically less than 2 to 4% of the soil, so it is dwarfed by the mineral fraction. However, a small amount of organic matter can significantly alter total CEC. Nitrogen, phosphorus, and sulfur are not dependent on CEC because they are contained in significant amounts in soil organic matter. They are not exchangeable, however, and must be mineralized through microbial action to become available.

SUSTAINABLE SOIL MANAGEMENT

The benefits of maintaining or enhancing soil structure through growing cover crops and/or compost has been discussed thoroughly in the previous section. Sustainable soil management does not end there, however. It means knowing what soil types you have in your vineyard and the properties of these soils such as moisture holding capacity, texture and rooting depth. You should sample the soil periodically to track any changes in soil chemistry. (See Table 11-1 for suggestions about taking a soil sample.) The soil samples should be coupled with vine tissue samples to monitor the levels of important nutrients. If you grow more than one or two tons of winegrapes per acre, a significant amount of nutrients are exported off the vineyard in the grapes after harvest and are not replaced. At some point even on the most fertile soils some of these nutrients will need to be replaced with additions of fertilizer from off site, either in the form of organic additions such as compost, or in synthetic fertilizers. In either case it is important to monitor the soil and the vines to determine the vineyard needs. The most accurate way to do this is to create a nutrient budget for the vineyard based on all of the factors listed above. A detailed guide for creating a nutrient management budget for your vineyard is presented in Chapter 14.

Table 11-1   Guide for taking soil samples.

- A soil sample should include at least 15–20 cores from a 20–40 acre block. This is a sufficient number of soil cores on a uniform field.

- If the field has soil type change or topographical changes (low and high points or is located on a slope), a sampling protocol should be devised to collect separate soil samples from the unique areas.

- Remove plant residues on the soil surface before taking the core sample.

- Cores should be taken from a depth of 12–18" for a normal sample, or to 18", 36", and 64" in certain circumstances to assess movement of fertilizer or to diagnose a problem or in developing a vineyard. Take the samples where the majority of the roots are located (e.g., under the drip emitters or under the furrows depending on irrigation system).

- To monitor nitrogen acidification or fertilizer salinization under drip emitters, take core samples directly under the emitters. To determine whether nitrogen is required for cover crops take surface soil (4" to 6") from mid row as described above. Mix the cores together well in a bucket making sure large clods and aggregates are completely broken apart. Randomly remove soil from the bucket by taking a number of scoops from different areas of the bucket to make a one pound (three cups) sub-sample for lab analysis.

- If the soil changes in the block, a separate sample should be taken from each area. The sample should be kept cool (minimum under shade; optimum refrigerated) and mailed ASAP to the soil test lab.

- If the whole block is to be treated the same, make sure that the proportion of different soil types in the sample is representative of the proportion of different soil types in the block.

There are other factors to consider when managing your vineyard soil sustainably. Soil compaction is something to be avoided. Maintaining good levels of organic matter through compost addition, reduced or no tillage, and cultivation of cover crops will help maintain soil aggregates, which reduces compaction. Avoiding equipment use on water-saturated soil is important. Using the lightest possible equipment, bigger diameter tires or tracked equipment helps to minimize compaction. The closer the tractor wheels come to the vine in a drip irrigated vineyard, the more compaction you will get in the root zone, so using as narrow a tractor as feasible is another way to reduce compaction. One way to keep the tractor wheel as far away from the vine as possible is to use over the row equipment. The wheels go down the middle of the vineyard alley. Moreover, if you use multi-row sprayers the tractor does not go down every row.

Soil erosion is a very important issue in sustainable winegrowing. It can occur through the action of wind or water and can happen to soil in the vineyard as well as on dirt roads or headlands. Wind erosion causes two problems. One is the loss of soil and the other is air pollution due to the suspension of particulate matter in the air. Dust in the air is a particular problem in more arid regions or in Mediterranean climates like the Central Valley of California where it does not rain during the growing season. You may have seen the terms PM10 (or PM5) in relation to air pollution. PM stands for particulate matter; PM10 means particulate matter less than 10 microns in diameter. These sizes are important because when particles less than 10 microns in diameter are inhaled and enter the lungs, the body cannot get rid of them. The particles are too small and remain in the lungs, creating health problems if enough accumulate. PM10s, therefore, are important air quality measures. PM10s originate from various sources, such as burning of diesel fuel, burning wood, and from dust created during agricultural activities. They are the focus of many air quality controls in California. It is important for winegrape growers to be aware of them and how to reduce their production.

Whether caused by wind or water, erosion it is prevented by anchoring the soil to prevent soil particles from moving off site. There are several ways to anchor soil to reduce erosion. The most sustainable one is to plant cover crops whose roots anchor the soil. Cover crops can be planted not only in the vineyard but also on headlands, and even on roads that do not sustain frequent traffic. Of course

Figure 11-4   Soil erosion in the vine row.

picking the right species of cover crop to plant is critical for optimum performance (Ingels et al. 1998). On most vineyard roads that sustain a significant amount of traffic the most efficacious form of anchoring the soil is by applying some type of dust suppressant. Some are more environmentally sound than others and their costs vary.

The San Joaquin Valley Air Pollution Control District[1] and the California office of the NRCS[2] suggest the following dust suppression materials:

- *Paving*—concrete or asphalt.
- *Oils*—bituminous/road oil (SC-80, SC-250, SC-350, SC-800, oil sand).
- *Roadmix*—crude-oil-containing soil mixed with aggregates and soils.
- *Road base*—standard road base meeting class II ¾" standards, crushed gravel meeting class II ¾" standards, ground-up asphalt.
- *Chemical dust suppressants or polymers*—calcium chloride, magnesium chloride, lignosulfate, calcium lignosulfate, petroleum emulsions, polymer emulsions.
- *Organic materials, chips/mulch*—materials such as almond shells or hulls, pistachio shells, chipped prunings, or other organic/vegetative materials placed on the road for dust suppression.

REDUCING EROSION AND SEDIMENT TRANSPORT FROM VINEYARD ROADS

Vineyard roads can be a major source of sediment pollution to streams – delivering damaging nutrient loads, smothering fish eggs, and reducing the variability in stream habitats (which, in turn, can reduce the number of plant and animal species a stream can support). It is important, therefore, to limit erosion from vineyard roads, and prevent erosion that does occur from reaching streams and other water bodies. Important road-related sediment reduction measures include:

Outsloping Unpaved Roads—Because roadbed erosion can only be completely abated through paving, management of unpaved roads should focus both on reducing erosion rates and preventing sediment that does erode from leaving the vineyard. Like insloping, outsloping roads minimizes surface erosion by rapidly moving water from the roadbed. However, outsloping has the added

1. *www.valleyair.org*
2. *www.ca.nrcs.usda.gov/news/publications/*

benefit of dispersing eroded sediments along the hill-slope (where it can be filtered out by cover crops or natural vegetation), rather than concentrating sediment in the ditch (where it can be delivered to nearby water bodies). In addition, by reducing or eliminating the need for ditches, outsloped roads are among the least expensive road types to build and maintain.

Vegetating Unpaved Roads—Planting unpaved surfaces in or around vineyards with grass or other vegetation (where feasible) can be a reasonable solution for reducing erosion and dust. (See Air Quality chapter for more detail on dust mitigation.)

Grassing and Hardening Ditches—Depending on the degree of slope, ditches should be grassed or hardened to prevent erosion. For low to moderate slopes, perennial grasses can be used to stabilize ditch surfaces and filter sediments from unpaved road surfaces. For steeper slopes and points of potential high scour, hardening ditch surfaces (e.g., with stone and/or cement) may be necessary to prevent ditch erosion and downcutting.

Stabilizing Culverts—Sediment erosion can occur at both the culvert inlet and outlet. At the inlet, culverts (especially if they are undersized) can impede the free flow of water and result in upstream erosion, often taking the form of an upstream "scour hole." At the outlet, concentrated flows can lead to downcutting and the development of a "perched" or "hanging" culvert, which in turn can cause even greater erosion of the downstream slope as water falls farther from the culvert outlet. To stabilize culvert openings, soil around inlets and outlets should be well compacted and points of scour hardened (e.g., with stone and/or cement). In addition, culverts should be sized to accommodate high flow events and installed at a slope matching the downstream grade.

Some soil management and erosion control practices can be expensive. The USDA Natural Resources Conservation Service (NRCS) has cost sharing programs that many growers may qualify for that will provide financial assistance to carry out some of these practices. Contact your local NRCS office in your county to see if there is a cost sharing program that will fit your needs.

REFERENCES

Horwath, W., C. P. Ohmart and C. P. Storm. 2008. "Soil management." *In* Ohmart, C. P., C. P. Storm, and S. K. Matthiasson. *Lodi Winegrower's Workbook.* 2nd Edition. Lodi Winegrape Commission, Lodi CA. 111pp.

Ingels, C. A., R. L. Bugg, G. T. McGourty, and L. P. Christensen. 1998. *Cover cropping in vineyards.* Univ. Calif. Div. Agric. Nat. Res. Publ. 3338. 162pp.

Ohmart, C. P. and S. K. Matthiasson. 2000. *Lodi Winegrower's Workbook: A Self-Assessment of Integrated Farming Practices.* Lodi-Woodbridge Winegrape Commission, Lodi, CA. 145pp.

White, R. E. 2009. *Understanding Vineyard Soil.* Oxford University Press, New York. 230pp.

12

Sustainable Water Management

Total annual rainfall and how it is distributed throughout the year varies greatly from one wine region to another around the US. In some areas of the country, such as parts of the Midwest and Eastern US, it rains throughout the year. California, where more than 90% of the winegrapes in the US are grown, has a Mediterranean climate where almost no rain falls between May and September. The grape-growing regions of Washington have very low annual rainfall and the growing season in the winegrape growing regions of Oregon are also dry.

The amount of rainfall and its seasonal pattern have both economic and environmental ramifications for vineyard management. For example, because of the presence of moisture and humidity during the growing season in the Midwest and Eastern US, disease pressure is much higher than it is in Mediterranean climates and drier regions, resulting in the need for more fungicide applications. Moreover, some diseases like Black Rot and Downy Mildew do not even occur in California vineyards. In low rainfall areas irrigation is necessary to get an economically viable yield. In regions where rainfall during the growing season is low or non-existent, irrigation management can significantly influence winegrape quality. In fact, in regions such as Lodi, one can argue that drip irrigation and how it is managed is possibly the single most important factor in regulating winegrape quality (Ohmart and Matthiasson 2000). Even in the Northeast, irrigation has been shown to be beneficial to quality, especially on gravelly soils that area shallow and have low water holding capacity (Tim Martinson, pers. comm.).

In irrigated vineyards, sustainable water management is not only concerned with optimizing water use, but also about providing enough water at the right times to produce the highest quality grapes possible. Therefore irrigation system design and engineering is critical. Optimizing water use involves practices such as monitoring the irrigation system to ensure optimum performance, monitoring ambient temperature to predict vine water use, assessing the level of vine moisture stress, and measuring the amount of soil water available to the vine. Timing when and how long to irrigate involves implementing practices based on research results, weather conditions, and experience with the vineyard soil, rootstock, and clone (Prichard et al. 2008).

Soil Water Holding Capacity

When determining irrigation amounts and frequency it is important to understand the role of soil water holding capacity. Water in the soil occupies the soil pores and spaces in between soil particles. The largest pores allow water to pass through, filling smaller pores. After irrigation or a rain, gravitational forces pull water out of the large pores, leaving water held in the soil only by capillary forces in the small pores. Clayey soils have very small pores and will hold more water per unit volume than sandier soils which have larger pores.

Following a complete soil wetting and subsequent de-watering of the large pores, a typical soil will have about 50% of the pore space filled with water and 50% with air. This condition is called the field capacity or "the full point." Soils begin to dry from vine water use and evaporation to a point where water becomes too difficult for roots to extract ("dry point"). Available water holding capacity is the difference between the wet point and dry point. The water available to the entire root zone is determined by multiplying the soil moisture (inches of water per foot of soil) by the root zone depth. Figure 12-1 and Table 12-1 illustrate the relationship between soil texture and available water content. When vines are using the first half of the available water, no vine water stress is expected. During the use of the last half, water stress does occur and increases as the dry point is approached. Available water increases as soil texture moves from sandy towards clayey textures and availability is maximum at a silt loam texture. As soils further increase in clay content, they hold more water but less of it is available to the vine due to clay's ability to hold water more tightly than coarser textures (Prichard et al. 2008). Therefore soil water holding capacity should be accounted for in a water management plan.

Vine Water Use

Before discussing vine water use, it is a good idea to briefly review growth of the vine and fruit during the growing season. The growth of shoots and leaves begins

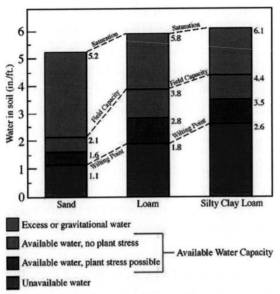

Figure 12-1 Relationship between soil texture and water content (Prichard et al. 2008). *Courtesy of Lodi Winegrape Commission.*

Table 12-1 Relationship between soil texture and water content (Prichard et al. 2008).

Available Water Capacity by Soil Texture	
Textural Class	**Available Water Capacity (Inches/Foot of Depth)**
Coarse sand	0.25–0.75
Fine sand	0.75–1.00
Loamy sand	1.10–1.20
Sandy loam	1.25–1.40
Fine sandy loam	1.50–2.00
Silt loam	2.00–2.50
Silty clay loam	1.80–2.00
Silty clay	1.50–1.70
Clay	1.20–1.50

shortly after budbreak (Figure 12-2). It proceeds at a high rate initially and then declines to near zero as veraison approaches. Nearly one half of the shoot length has occurred by flowering. Berry growth rate increases after flowering in an initial rapid period of growth (Stage I). In the second stage of berry development, growth is much slower, followed by another rapid growth rate near veraison (Stage III). Vegetative growth rate of the shoot continues to decline during Stage I of berry growth and by Stage III has pretty much stopped. Root growth, measured as the number of actively growing root tips per square meter of soil, has two distinctive high growth rate periods—one at flowering and another near and post-harvest (Figure 12-2). Berries begin to ripen at veraison. They begin to soften at this stage, change color, and begin to accelerate in growth during this third and final stage of growth (Stage III) (Prichard et al. 2008).

Knowledge of these growth patterns is important if one is interested in using irrigation to manipulate growth of the vine canopy and fruit. Water is essential for vine and fruit growth because it is required for cell expansion. In climates, such as eastern Washington, where rainfall is very low (less than 5 inches per year), one can significantly affect vine and fruit growth using irrigation, because virtually all of the water for the vine is being provided by the irrigation. In climates where rainfall is much higher, for example 15 inches or more per year, the timing of the rainfall and the amount can have a significant impact on vine and fruit growth, giving the grower much less control using irrigation. Therefore using irrigation in vineyard

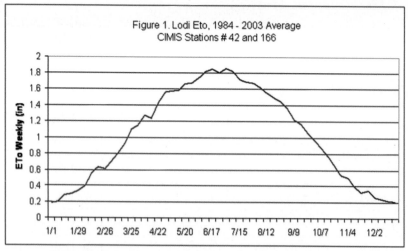

Figure 12-2 Growth of shoot, roots and berries (Prichard et al. 2008).

management is all about understanding the growth stages of the vine and fruit as well as the effects of water on this growth.

Water used by the vine is absorbed through the roots and moved into the vine. Water uptake is driven by the tension that is created within the vine's water conducting vessels as a result of water being released into the atmosphere through stomates in the leaves. Evapotranspiration (ET) is a measure of the amount of water evaporated from the soil and that used by a plant, and is a function of solar radiation, temperature, humidity, and wind speed. If you grow grapes in California, ET can be obtained from weather stations operated by the California Irrigation Management Information System *(www.cimis.water.ca.gov)*. It is likely many other states have ET measurements available through university or government services. A quick search on the Internet will provide links to these data. The figure obtained from a CIMIS station in California is a measure of the amount of water used by a unit area of mowed grass and it is termed the Evapotranspiration Reference Value or ETo. As one would expect, as temperature increases and then declines during the growing season ETo does the same (Figure 12-3).

Water use by the vine is also influenced by the amount of vine canopy present. To obtain an estimate of the full water use by a grapevine under no moisture stress, the CIMIS ETo measure is multiplied by a crop coefficient, Kc, for grapes that was determined through research (Prichard et al. 2008). Kc increases as the leaf area expands in the canopy from zero at bud break to a maximum at full canopy.

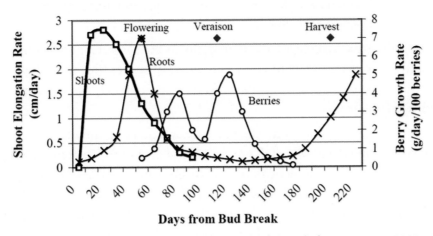

Figure 12-3 Average ETo's throughout the year for Lodi from 1984 to 2003.

The most accurate way to determine the amount of water to apply to your vineyard is to determine the water demand for the vines using ETo and the Kc. One cannot predict the amount of water used by the vine because one cannot accurately predict the weather. However, by using recent data to calculate the vine water use, such as over the previous week, one can irrigate to replace the amount of water used.

Drip Irrigation and Wine Quality

Drip (or "low volume") irrigation has revolutionized viticulture in arid production regions like certain parts of California (Figure 12-4) and has had significant impacts in higher rainfall regions like New York (Tim Martinson, pers. comm.). Low volume irrigation systems allow small amounts of water to be applied slowly and frequently through emitters spaced along polyethylene tubing. Water is uniformly applied to the vineyard. With properly installed and maintained low volume irrigation systems, vineyards are more uniform, healthier, and make better wine. Unfortunately, many of these highly engineered systems are not managed in a manner to realize their full potential. The extensive self-assessments done in the California wine industry under the auspices of the California Sustainable Winegrowing Alliance showed that drip irrigation systems were not maintained to the level necessary to ensure optimum performance (CSWA 2004). There is a need for constant monitoring and maintenance so problems such as clogged emitters or leaks can be repaired and distribution uniformity can be corrected to ensure optimum system performance (Prichard et al. 2008).

One of the great advantages of drip irrigation is the control it provides in deciding exactly how much water to apply and when. However, this flexibility brings with it the responsibility to schedule irrigations appropriately for managing vine growth and fruit quality.

In years past irrigations were scheduled to apply an optimum quantity of water to maximize productivity (i.e., full vine water use or more). However, it has long been recognized that water deficits created at the right time can lead to improved fruit quality—especially in red wine varieties (Prichard et al. 2004). Typical strategies to achieve water deficits were developed using surface irrigation and relied on irrigation cutoff to limit water as the fruit ripened. This resulted in both successes

Figure 12-4   Drip irrigation in a young vineyard.
Courtesy of Lodi Winegrape Commission.

and failures, depending on the timing of the cutoff and climatic and soil storage conditions (Prichard et al. 2008).

In the late 1990s, it became clear that maintaining a moderate vine water deficit can improve the partitioning of carbohydrate to the grapes while controlling excessive vegetative growth, giving rise to an irrigation strategy termed Regulated Deficit Irrigation or RDI. By restricting irrigation water volumes, soil water available to the vine becomes limited to a level where transpiration of water from the leaves exceeds water absorption from the soil by the roots. It is at this point that the vine begins to undergo a water deficit. RDI can be a consistent reduction (i.e., consistent reduction of planned irrigation volumes over the entire season) or a variable reduction over the irrigation season to induce the desired vine response at the appropriate time. The timing of the reduction has various effects. For example, deficit irrigation between berry set and veraison reduces berry size by reducing the number of cells in the berry. It also can reduce vegetative canopy growth exposing the fruit to more sunlight. Post veraison deficit irrigation can reduce berry size by reducing the cell expansion in the berry and can reduce the number of leaves in the canopy (Prichard et al. 2004). If one is to practice deficit irrigation it is very important to pay attention to vine water demand so that the vines are not overly stressed. Not only does properly applied RDI improve winegrape quality, it also reduces overall water use in the vineyard.

Successfully applying RDI to a vineyard requires accurate soil moisture or plant "stress" sensing, the ability to estimate crop demand, and the capacity to irrigate frequently using low-volume irrigation. It also requires experience with a vineyard as to how it responds to the heat waves that often occur in parts of California. A disadvantage of RDI is that it requires water status to be maintained accurately within a rather narrow tolerance; any excess irrigation reduces the advantage of the regulated deficit and can cost more in terms of water used, while under-irrigation can lead to yield or quality losses as a result of reduced photosynthesis, excess fruit exposure and/or fruit shrivel near harvest (Figure 12-5) (Prichard et al. 2008).

Like any new vineyard practice that shows promise in lowering vineyard inputs while increasing wine quality, RDI was adopted quickly by many growers. Some concluded that if a slight to moderate RDI improves quality, then a more severe RDI would improve the quality even more. However, some people overdid it, resulting in yield loss and crop damage. Some growers overreacted to these experiences and concluded that RDI does not work. The reaction was strong enough that

Figure 12-5   Severe vine moisture stress due to inadequate irrigation.

these growers will not even use the term RDI but prefer "irrigation management. " Nevertheless, RDI is a very important practice and like any new approach requires some getting used to.

Developing a Water Management Plan

Water management in irrigated vineyards has to be tailored to each site. How much water is applied and the timing of it depends on many things like soil type, variety, clone, rootstock, and the client who is buying the grapes. Therefore, sustainable approaches to water management will differ with each vineyard. However, as with many other aspects of sustainable winegrowing, much of sustainable water management can be achieved by attention to detail through developing, writing down, and executing a water management plan. The *Lodi Winegrower's Workbook* (Ohmart et al. 2008) provides guide to this approach and is the source of most of the information presented below.

A water management plan should be designed to address the specific vineyard challenges, executed, and then monitored to determine its effect. Adjustments can be made in succeeding years to fine tune the plan to achieve one's goals.

For newly planted irrigated vineyards, it is important that a good root system be established to provide a foundation for a healthy vineyard throughout its life. Therefore a newly planted, non-bearing vineyard should be irrigated at an amount and scheduling to provide full water use for several seasons until it has reached full production.

In a full production vineyard, a plan to control excessive vine vigor and open up the canopy so more diffuse light penetrates to the fruiting zone generally requires that no irrigation be applied until shoot growth is under control as a result of reduced water availability in the root zone. Using visual clues or measuring leaf water potential are successful methods for allowing the vines to experience moderate water deficits before irrigating them. Also, measuring water potential can indicate when water stress is too high so irrigation can be applied to avoid canopy/fruit damage.

Moderate levels of vine water stress vary somewhat by variety and fruit quality goals. White winegrapes can benefit from reduced vegetative growth through RDI. However, the resulting open canopy and excessive light exposure to the fruit may create more color and character than desired or even sun-burn. Mid-day leaf water potential of –10 to –13 bars is a typical irrigation start point or "Leaf Water Potential Threshold" for a white winegrape RDI.

Red winegrapes on the other hand, tend to benefit from higher levels of stress. Red winegrapes develop desired fruit characteristics with a more severe threshold of –13 to –15 bars. Some red grape varieties are more sensitive than others, with Merlot being most sensitive and varieties like Cabernet Sauvignon, Syrah, and Zinfandel exhibiting more tolerance.

Once an irrigation "threshold" is selected, the question becomes how much water to apply (a specific RDI %). The RDI is a percentage of the full amount of water that a vine will potentially use, based on climatic demand and canopy size when water from the soil is not limited. The selected RDI should prevent shoot growth from resuming but allow photosynthesis to continue, prevent excessive leaf loss, and limit fruit sun-burning or raisining.

Successful RDI's are typically 50 to 60% of full vine water use. More severe RDI's of 35–40% have resulted in delayed harvest and poor quality fruit in seasons when harvest is late. Deep root zones and clayey soils can sustain vines at lower RDI's. Care needs to also be taken if extended maturity harvests are practiced (i.e., long hang time).

Deficit Irrigation (RDI) is practiced to limit excessive vegetative growth and improve winegrape quality or limit water use in times of drought. However, RDI is not appropriate for all situations. Examples where a full water regime is appropriate include young developing vineyards, low vigor vineyards resulting from particular rootstock/scion combinations, vineyards on soils with poor water holding capacity, or vineyards with nutrition or pest related issues.

A sustainable water management plan will consist of assessing the quality and amount of water available to the vines from the soil, determining how much water the vines need over and above what is available to them to obtain the desired yield and quality, and delivering that amount of water to the vines in the optimum manner to minimize losses from the vineyard system. All of these actions involve monitoring. Remember the adage from Chapter 10: "If you cannot measure it you cannot manage it." The same holds true here. I will not go into the details of each component of the water management plan because that can be obtained elsewhere (e.g., Prichard et al. 2008). I will briefly comment on each one to remind the reader of important things to consider.

MONITORING WATER QUALITY

Irrigation water can contain a range of contaminants depending on its source, such as boron, iron, sodium, chlorides, calcium, nitrates, suspended solids, and bicarbonates. It is important to know the quality of the water you are putting on your vineyard so that it can be treated if necessary to avoid situations such as salt buildups, irrigation system clogging, mineral toxicity to the vines (e.g., boron), and water penetration problems. Moreover, some water contains nitrates at high enough levels such that several pounds of nitrogen per acre are applied with the irrigation water during the growing season. This could significantly affect your vineyard nitrogen management program. I do not think there is an agreement on how often irrigation water should be tested. In developing the *Lodi Winegrower's Workbook*, the workbook committee felt that three to five years was adequate (Prichard et al. 2008). In California, well water should be tested more often than irrigation water from canals or other surface sources. Be sure to research any regional water quality issues so you are aware of problems before they impact your vineyard.

SOIL AND PLANT MONITORING

There are many ways to monitor soil moisture, such as with neutron probes (Figure 12-6), gypsum blocks, tensiometers, and the old fashion "feel" method using a bucket auger. Each method has its advantages and disadvantages. In some regions companies will, for a price, set up soil moisture monitoring systems in your vineyard. Your goals and your budget will help you determine which one is right for your vineyard. One thing that is certain is that if you choose to deficit-irrigate your vineyard you will need a quantitative measure of soil moisture.

There are not as many ways available to monitor moisture stress in the vine. The old-fashioned way is to use visual symptoms as a diagnostic tool. Table 12-2 provides an example for visual assessments of shoot tips, leaves, and fruit. For the last few years some growers in California have been using pressure "bombs" that directly measure moisture stress in leaves. This method is used primarily to decide when to start the first irrigation of the year. Some viticulturists feel that measuring moisture stress in this way is the only measurement needed to assess vine water stress. They argue that this is a direct measure whereas measuring soil moisture is an indirect measure of what is happening in the plant. Others feel it is important to use several measures to keep track of vine moisture status and irrigation require-

Figure 12-6 Neutron probe being used to measure soil moisture. *Courtesy of Lodi Winegrape Commission.*

ments, such as soil moisture levels, plant moisture stress, evapotranspiration, and visual symptoms.

There have been several excellent, detailed descriptions of how to calculate irrigation scheduling for California vineyards such as Prichard et al. (2004)[1] and Prichard et al. 2008. I will not repeat the information here but recommend you use these references to help you create a quantitatively-based water management plan for your vineyard.

IRRIGATION PUMPS

Water is pumped to the vineyard using either an electrical motor or a diesel engine. What type one chooses will most likely depend a great deal on economics, but there are also environmental and regulatory issues to consider, at least in California.

Diesel engines are classified according to Tiers. A Tier 2 diesel pump has a diesel engine that has a specific level of exhaust emissions for things such as particulate matter (PM), carbon monoxide and nitrous oxides (NOx). The first federal standards (Tier 1) for new non-road (or off-road) diesel engines were adopted in 1994 for engines over 37 kW (50 hp), to be phased in from 1996 to 2000. In 1996, a Statement of Principles (SOP) pertaining to non-road diesel engines was signed

---

1. For a free copy visit *www.lodiwine.com/Final_Handbook.pdf*

Table 12-2 Qualitative indicators of vine moisture status.
(Each series of steps goes from no stress to severe stress.)

Shoot Tips
Shoot tips actively grow beyond the expanding leaves. Tendrils aggressively reach above the growing tip. Leaves stand perpendicular to the sun's rays and show bright green coloration.
Shoot tips are actively growing. The last expanded leaf expands behind the tip. Tendrils are even with the growing tip and basal tendrils droop. Leaves hang perpendicular to the sun's rays.
Shoot tips grow less actively. The last expanded leaf folds over the tip. Tendrils dry and detach at the base of the cane if not fastened to something. Tendrils near the tip droop. Leaves begin to bend and shy away from the sun's rays and change color to duller shades of green.
Growing tips begin to yellow or brown and are clearly not growing. The last expanded leaf folds over and shields the tip. Tendrils near the tip droop or easily fall off when touched. More leaves bend away from perpendicular and now hang more parallel to the sun's rays.
Growing tip dies or falls off. All non-attached tendrils dry and fall. Leaves yellow and drop due to stress. The balance of leaves darken and thicken.

Leaves
No leaf loss due to moisture stress.
2–10 leaves lost or yellowed per vine, leaf color changing to dull green.
10–30 leaves lost or yellow per vine.
Leaf loss up to and including the fruit zone.
Leaf loss above the fruit zone.

Fruits
No signs of fruit dehydration, clusters feel firm to the touch.
Some noticeable signs of fruit dehydration, clusters are softening to touch.
Greater than 5% of the vines with some fruit dehydration, evident berry changes are visible.
Less than 40% of the vines show signs of some fruit dehydration.
Greater than 40% of the vines show signs of some fruit dehydration.

between EPA, California Air Resources Board and engine makers. On August 27, 1998, the EPA signed the final rule reflecting the provisions of the SOP. The 1998 regulation introduced Tier 1 standards for equipment under 37 kW (50 hp) and increasingly more stringent Tier 2 and Tier 3 standards for all equipment with phase-in schedules from 2000 to 2008. The Tier 1-3 standards are met through advanced engine design, with no or only limited use of exhaust gas after treatment (oxidation catalysts).[2]

On May 11, 2004, the EPA signed the final rule introducing Tier 4 emission standards, which are to be phased in over the period of 2008-2015. The Tier 4 standards require that emissions of PM and NOx be further reduced by about 90%. Such emission reductions can be achieved through the use of control technologies—including advanced exhaust gas after treatment—similar to those required by the 2007–2010 standards for highway engines.

It is clear from the above discussion that the type of diesel engine one uses in the vineyard is regulated by state and federal laws.

When using an electrical pump one should consider obtaining the electricity from a renewable energy source. More and more growers are installing solar arrays in the vineyard to provide electricity for the pump.

Photovoltaic solar panels capture the light energy of the sun and convert it to direct current power (Figure 12-7). When connected to the electric utilities power grid and time-of-use metering, a solar system can capture energy during the daylight hours sending the power to the grid at the peak hourly rate and retrieving energy from the grid, to run an irrigation pump during the night at the off-peak rate.

A solar system is effectively delivering power to the grid for a higher price than an irrigation pump running at night is drawing out of the grid. This advantage helps recoup some of the costs of installing the system. However, it should be noted that in some regions of California the current system total credit cannot exceed actual use for the day or season. A net metering time-of-use system keeps track of the power produced by the solar panel and banks that power for when the energy is later drawn. Another advantage of a system like this is that power is produced year round, so all the energy captured on the 300+ sunny days a year in Lodi can be drawn when needed at no cost.

---

2. Source: *www.dieselnet.com/standards/us/offroad.html*

Figure 12-7 Irrigation pump with solar array to power it.

Unfortunately, because a grid-connected system is just that, connected to the grid, when a power outage occurs, energy is not delivered. Alternatively, an off-the-grid system has the advantage of being able to supply power during a power outage from a large number of batteries, but the cost is significantly higher. For relatively small systems, like those used to run an irrigation pump, a time-of-use grid-connected system has the most advantages.

Once the type of pump has been selected there are other issues to consider, such as its efficiency of operation. Field testing programs have shown that overall efficiencies for electrically driven pumps average less than 50 percent, as compared to a realistically achievable efficiency of at least 67 percent. This implies that 25 percent of the electrical energy used for pumping is wasted, resulting from poor pumping efficiencies alone. Depending on acreage and water volume pumped, winegrowers could significantly reduce energy costs per well by increasing pumping plant efficiencies from present average levels to higher levels (Prichard et al. 2008).

There are several reasons that pumps become less efficient with use, such as wear due to sand in the water, clogged impeller, improperly matched pump, poor suction conditions and changes in ground water levels. Ideally, pump efficiency should be checked every three years (Prichard et al. 2008). Some regional power companies will cost-share a pump test.

IRRIGATION SYSTEM DISTRIBUTION UNIFORMITY

Drip irrigation systems have revolutionized how growers irrigate their vineyards, but they do not check the performance of these systems often enough. Most growers will check the system for leaks frequently, often during every irrigation. However, checking for leaks does not tell you how evenly it is distributing the water in the vineyard. If a grower is willing to put out the money for an expensive drip irrigation system, time and effort should be put into making sure it is performing at its peak. This is very much a part of sustainable winegrowing. Prichard et al. (2004) present a succinct description of how to check distribution uniformity.

DRIP IRRIGATION SYSTEM MAINTENANCE

Drip irrigation systems may require a significant amount of maintenance to continue operating at maximum efficiency. If you discover that your distribution uniformity is not what it should be, part of the problem might be rectified by system maintenance. Routine maintenance is very much a part of sustainable water management and can include checking for leaks, back-flushing filters, periodically flushing lines, chlorinating, acidifying, and cleaning or replacing clogged emitters.[3]

Flushing the System

Filters should be back-flushed periodically to clear any collected particulate or organic matter. Clogged filters can reduce pressure to the system, lowering the water application rate. Back-flushing can be done either manually or automatically.

Mainlines, sub-mains, and particularly lateral lines should be flushed periodically to clear away any accumulated particulates. Mainlines and sub-mains are flushed by opening the flush valves built into the system for that purpose. When

---

3. Source: Prichard et al. 2008

Figure 12-8   Sand media irrigation system filters.
Courtesy of Lodi Winegrape Commission.

the system is designed, the flush valves should be sized large enough to allow the water velocity to move particulates out of the system.

Polyethylene lateral lines are flushed by opening the lines and allowing them to clear (Figure 12-9). This measure is essential since the filters trap only the large contaminants entering the system, allowing the lateral lines to collect material that may eventually clog the emitters. Flushing clears the system of many contaminants.

Chlorinating

Water with a high organic load (algae, moss, and bacterial slimes) should be chlorinated with chlorine gas, sodium hypochlorite, or calcium hypochlorite. Whether to chlorinate continuously (1 to 2 ppm free chlorine at the end of lateral line) or periodically (approximately 10 ppm free chlorine at the lateral end) depends on the severity of the clogging. Continuous chlorination is usually necessary when the clogging potential is severe. Surface water is more likely than groundwater to cause

Figure 12-9   Flushing an irrigation line.

organic clogging. Well water pumped into and stored in a pond or reservoir should be considered a surface water source.

Acidifying

Acidification may be required for irrigation water that has a tendency to form chemical precipitates such as calcium carbonate (lime) or iron. Groundwater sources are most susceptible to chemical precipitation.

Lowering the pH of the water to 7 or below is usually sufficient to minimize calcium carbonate or lime precipitate problems. Acids that can be added to the irrigation water include sulfuric, hydrochloric or murietic acid, and phosphoric acid. A urea-sulfuric acid fertilizer is frequently used and is safer to handle. Acidification has the added benefit of increasing the efficacy of chlorine additions, but the acid and chlorine sources should never be mixed together, since hazardous chlorine gas will be formed.

Cleaning or Replacing Emitters

Emitters used for permanent crops may have to be cleaned or replaced because of clogging. The laterals and emitters should be inspected routinely to identify drip emitters that are completely clogged, although this will probably not identify emitters for which the flow has been only reduced. Partially clogged emitters can be located by collecting water from the emitters to determine their discharge rate. The first step in cleaning emitters is determining what has caused the clogging. Material caught when the laterals are flushed can be examined. If organic matter is found, a high level of chlorine (approximately 50 ppm) can be injected into the line for 2 hours or more and then allowed to sit for about 24 hours. The lines should then be thoroughly flushed. If chemical precipitation appears to be the cause, then acid can be injected for 2 hours or more to lower the pH to approximately 5. The acid should be allowed to sit in the line for 24 hours and then flushed. If injecting chlorine or acid doesn"t clear the clogging, the faulty emitters may have to be replaced, in which case it is usually wise to leave a clogged emitter in the line and simply to install a new emitter nearby. Although some brands of drip emitters can be disassembled and cleaned, nearly all are permanently sealed. In the case of pressure compensating emitters, contact the manufacturer to determine the minimum pH allowable. The flexible orifices in some pressure compensating emitters are damaged by water with a pH of 4 or below.

Inspecting or Replacing Other Parts

Other maintenance tasks to be carried out on a less frequent basis include inspecting the filter media, pressure-regulating valves, flush valves at the end of laterals, and replacing pressure gauges.

A sand media filter tends to cake together over time and as a result may fail to provide good filtration. Frequent back-flushing may be symptomatic of such a problem. The sand media should be replaced if this occurs. When the old media is removed, the underdrain system should be inspected. Even if the sand media appears to be in good condition, adding media periodically may be required, since some of the sand is invariably lost during the back-flush cycle.

The adjustable pressure-regulating valves, set at installation, should be inspected and adjusted periodically to see that the correct operating pressure is maintained. Preset pressure regulators should be inspected to ensure that they are

Figure 12-10 Irrigation lines with large leaks can result in significant water loss and under irrigation of vines farther up the line.

operating properly. Foreign material in the line may jam the adjustment mechanism and inhibit operation.

Pressure gauges tend to wear out eventually and should be replaced if their accuracy is in question. Liquid-filled pressure gauges, which are slightly more expensive than the standard gauges, may be a good replacement choice. The gauges must be scaled to operate in a pressure range appropriate for the system.

Micro-irrigation systems should be inspected regularly for leaks, a task that can be performed at the same time as checking for clogged emitters. Leaks can occur in hardware (compression fittings, end closures, emitter barbs, and hose adapters) or when the aboveground polyethylene tubing is damaged by farm equipment, harvest activity, or animals.

FERTIGATION

If you have a drip irrigation system in your vineyard, it provides you with the opportunity to apply fertilizers through it. This is called fertigation and is the common term for injecting fertilizers or soil amendments through the irrigation system. Applying these materials through a micro-irrigation system has many benefits:

- Makes fertilizer distribution relatively uniform
- Allows flexibility in timing fertilizer application
- Reduces labor required for application in relation to other methods

- Allows for reduced fertilizer application rates
- Reduces overall costs

However, water source contamination can occur during fertigation if the irrigation water pumping plant shuts down while the injecting equipment continues to operate, causing contamination of the water source or unnecessary amounts of fertilizer to be injected into the irrigation system. This can also happen if the injection equipment stops while the irrigation system continues to operate, causing the irrigation water to flow into the chemical supply tank, which can overflow.

Backflow prevention devices, which include check valves and vacuum breakers, are required on well heads to prevent water source contamination. If the injection pump is electrically driven, an interlock should be installed so that the injection pump will stop if the irrigation pump shuts down. To keep water from flowing backward into the chemical tank, a check valve or normally closed solenoid valve (with pump interlock) should be installed in the injection line.

IRRIGATION FLOW METERS

Irrigation flow meters work like the speedometer in your car, except, instead of measuring vehicle speed in miles per hour, they measure the current pumping rates, usually in gallons per minute. Flow meters are also like your car's odometer in that they track the total gallons or acre-inches pumped. Flow meters are an integral part of any irrigation-scheduling program. They are usually installed between the pump and the mainline (Figure 12-11). They can be used to confirm that the scheduled amount of water was delivered to the vineyard. The gallons per minute readout can help set diesel pumps for a specific flow rate and can help indicate the existence of leaks or emitter clogging.

In states like California where water is at a premium, there is likely to be a growing concern over how much water it takes to produce a unit of crop (i.e., the "water footprint"). In the case of winegrapes the water footprint will be how much water it takes to produce a ton of grapes. A flow meter will be able to provide the measure of how much water was applied per acre in order to make this very important measure of sustainability performance.

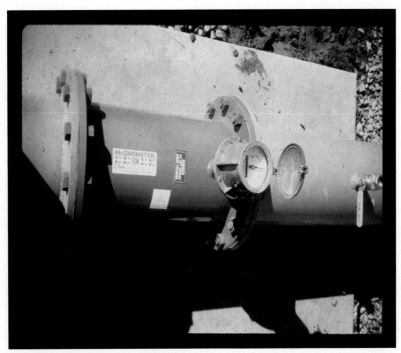

Figure 12-11 Irrigation flow meter.
Courtesy of Lodi Winegrape Commission.

REFERENCES

CSWA. 2004. *California Wine Community Sustainability Report. Chapter 5. Vineyard Water Management.* California Sustainable Winegrowing Alliance, San Francisco.

Prichard T., B, Hanson, L. Schwankl, P. S. Verdegaal, and R. Smith. 2004. Deficit Irrigation of Quality Winegrapes Using Micro-Irrigation Techniques. Univ. Calif. Coop. Ext. Dept. Land, Air, Water Res. Univ. Calif. Davis. 89pp.

Prichard, T., C. P. Storm, and C. P. Ohmart. 2008. "Water management." *In* Ohmart, C. P., C. P. Storm, and S. K. Matthiasson. *Lodi Winegrower's Workbook 2nd Edition.* Lodi Winegrape Commission, Lodi CA. pp.142–186.

Ohmart C. P., and S. Matthiasson. 2000. *Lodi Winegrower's Workbook: A Self-Assessment of Integrated Farming Practices.* Lodi-Woodbridge Winegrape Commission, Lodi CA. 145pp.

13

IPM, the Foundation of Sustainable Pest Management

Integrated pest management (IPM) should be a fundamental part of any sustainable winegrowing program. It is cost-effective, flexible, and resilient. Since its inception in 1959, IPM has evolved into the most sustainable approach to managing pest problems in the vineyard (Stern et al. 1959).

WHY DO WE HAVE PEST PROBLEMS?

When considering pest management, I think it is helpful to start by asking why we have so many pest problems. A grower who farms using organic or Biodynamic methods is likely to say it is because we use too many pesticides and have gotten away from the natural way of farming. While this is partly true it shows a lack of understanding of pest population dynamics.

One can describe pest population fluctuations in very simple terms. The number of pests present is the number of births minus the number of deaths plus the number of individuals immigrating into the area, which in the case of most pests is not a significant number. Pest numbers go up when the number of births exceeds the number of deaths and goes down when the number of deaths exceeds the number of births. Deaths can be caused by a number of things, but most often for insect pests it is due to natural enemies killing them. Biological control works because natural enemies kill pests at a faster rate than they are born. Other important mortality factors are pathogens of pests, extreme cold or heat, and starvation.

The role of birth rates of pests is often overlooked as a cause of pest outbreaks. The quality of food that a pest consumes can have a very significant effect on birth rate. Higher food quality allows the female to mature more quickly, grow larger, and lay more eggs. Moreover, they may live longer which may also allow them to lay more eggs (Ohmart et al. 1991). It turns out food quality is the most important factor in regulating the populations of some pest species while others are regulated by their natural enemies, and in some cases, both factors come into play. One of the outcomes of the role of food quality in insect population dynamics is that biological control is not an important control factor for all pest species because some are not regulated by natural enemies but by the quality of the food they eat.

In Chapter 2 I pointed out that even in the most "natural" farming systems the situation is very far from a truly natural system. Most of the crops we cultivate, including *Vitus vinifera*, are grown in regions in which they did not evolve. Furthermore, many of our most important pests are ones that did not evolve with the crop. Often in their native ranges they are not pests, but only become so when they get transported to new regions. For winegrapes, examples would be pests like the glassy-winged sharpshooter, variegated leafhopper, and the vine mealybug. The result is an unstable ecological system, exotic crops being attacked by exotic pests.

Additionally, the varieties of crops we grow are very different from the original species from which they evolved, as a result of our selecting for their most desirable traits. In most cases the plant parts we harvest to eat, such as foliage, fruits and seeds, are packed with nutrients, which not only make them desirable to us but also makes them desirable to insects and diseases (Figure 13-1). We do not eat plant stems, with the exceptions of a few species like celery. That is because they are full of chemically complex structural compounds and plant secondary chemicals that are hard to digest and/or toxic, not only for us but also for insects and diseases. Very few species of insects and pathogens attack plant stems, for the same reason we do not eat them. The end result is that our gardens and farms are full of plants containing parts that are highly attractive and nutritious for pests, creating an ecologically unstable situation despite our best intentions otherwise.

There are many results of these unnatural situations. Biological control agents may not be as effective as in a natural situation or food quality may be enhanced for some pest species. When we put all of the above factors together we begin to understand why we will always have pest problems. It is true that a more natural system will be more stable than a less natural system and on the whole have fewer pest problems. But even in the most natural farming systems, pest outbreaks will still occur off and on due to the factors mentioned above. Even relatively undisturbed forest systems sustain pest outbreaks due to either a temporary breakdown of

Figure 13-1   Zinfandel grapes infected with Botrytis.

biological control, a change in food quality, or a combination of both (Barbosa and Schultz 1987, Elliott et al. 1998).

It is worth noting that viticulturists in many areas in the Midwest and Eastern US are growing native and hybrid winegrape species that have much more resistance and/or tolerance to many of the native disease and insect pests. This is a much more sustainable approach to pest management, at least for native pests, than vineyards where non-native species, such as *Vitus vinifera* are grown.

The History of Pest Control

To understand why IPM is the best way to manage pest problems, as well as to understand why pesticides have had such a big influence in agriculture, both positive and negative, it is helpful to trace the history of pest management.

Synthetic pesticides are some of the most common pest management tools used today. However, it is important to realize that they have only been available since the 1940s. While this may seem like a long time ago to younger readers, it is very recent when looking at the long history of agriculture. It is worthwhile to consider what growers used to control pests before the development of these materials. Before the invention of synthetic pesticides, growers managed pest problems as best they could, many times with a combination of techniques. In some agricultural systems, crop rotation was important. A pest would build up on one crop but before it could become very damaging, another crop on which it could not survive was planted in its place, keeping the pest population at a manageable level. This, of course, was used primarily with annual crops. However, growers in California have noticed a significant reduction in the productivity of some orchard crops, such as almonds, when they are replanted on the same site as a previous orchard of the same species. This is called a replant problem, and the current thinking is that it is caused by the build-up of soil organisms that negatively affect the second generation orchard.

Another important pest management strategy used in some very specific situations prior to the invention of synthetic pesticides was the use of biological control. While some might think of biological control as a relatively new way of managing pests, the first documented case of a successful biological control dates back to the late 1800s in California citrus orchards. At that time the budding California citrus industry was threatened with disaster when the cottony cushion scale, *Icerya pur-*

Figure 13-2 *Icerya purchasi*, the cottony cushion scale.

chase, was accidentally introduced into California. C. V. Riley, the state entomologist, traveled to the scale's native home in Australia, identified a lady beetle natural predator known as the Vedalia beetle, brought it back to California and released a few hundred. What happened next is now biological control lore. The beetle multiplied quickly and brought the cottony cushion scale population under control. It has remained that way since then, unless something like the misuse of pesticides reduces the Vedalia beetle population to a level where cottony cushion scale is no longer controlled (Debach and Schlinger 1970).

Like any pest management strategy, biological control can have unintended consequences. Probably the most recent case related to winegrapes is the problem occurring as a result of the multi-colored Asian lady beetle entering grape clusters near harvest. The beetle was introduced in the Midwest to control soybean aphid. When soybeans are harvested, the beetles migrate to other areas, and later in the summer are attracted to potential overwintering sites. Some find their way into grape bunches. If they remain there after harvest and through crush, their body fluids cause wine taint, and it only takes a few individuals to make the wine not marketable.

Using varieties that are genetically resistant to important pests was another important pest management approach prior to the 1940s. One of the most well-known cases involves winegrapes. Grape phylloxera, *Daktulosphaira vitifoliae*, is an aphid-like insect that is native to the Mississippi Valley and eastern US. It has a complicated life cycle and one of its forms lives on roots of North American grape species (Figure 13-3). In the mid 1800s it spread from its native range to other parts of the US and France, where it threatened to decimate the French wine industry because *Vitis vinifera* was extremely susceptible to grape phylloxera. Scientists at the time observed that North American grape vines were not susceptible to phylloxera

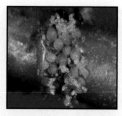

Figure 13-3   Grape phylloxera on a grape root.

and got the idea of grafting European grape scions to North American grape root-stock. The result was a plant resistant to the root feeding form of phylloxera. The French wine industry was saved by the use of these resistant rootstocks. These root-stocks are also the foundation of phylloxera pest management in most regions where the pest now exists and where *V. vinifera* is a critical part of the wine indus-try in the region.

Other pest management techniques used prior to synthetic pesticides include: tilling to kill insect and weed pests, changing planting dates to avoid peak pest numbers, and planting cover crops to provide refuge for pest natural enemies. All of these strategies are ones still used today.

THE DDT MIRACLE: THE ANSWER TO ALL OUR PROBLEMS

Up until the 1940s there were few pesticides available. One was Bordeaux mixture, a combination of copper hydroxide and hydrated lime that controls downy mildew. The story goes that its fungicidal properties were discovered by accident. In some places in France theft of winegrapes was a problem and supposedly someone used Bordeaux mixture to treat the grapes to prevent this theft. They observed that grapes treated in this way did not get downy mildew. Thus a fungicide was born. Other non-synthetic pesticides used prior to the 1940s are nicotine sulfate and ele-mental sulfur. Despite being naturally derived and approved for use on organically certified crops, nicotine sulfate is one of the most toxic compounds known to warm-blooded animals. So its use is problematic.

Pest management changed dramatically with the discovery of the insecticidal properties of nerve toxins developed and used during World War I against soldiers. The importance of synthetic pesticides really became evident with the discovery of the chlorinated hydrocarbons such as DDT (Figure 13-4). At the time, these mate-rials appeared to be the "silver bullet" for pest control for which everyone had been looking. They seemed perfect: they were cheap, effective in small amounts and

Figure 13-4 A can of DDT wettable powder.

against a broad range of insects, had long residual activity once applied, and unlike nicotine sulfate were not acutely toxic to animals and people. Initially, insect control was so effective that some entomologists predicted the eradication of entire species of pests. Demand for pesticides grew very fast. To meet this demand an immense industry arose to produce chemicals and to develop new, more effective ones.

At first, control was spectacular and before long pest control and chemical control became synonymous. At this point in time whenever a pest problem arose, only two questions were asked: 1) What pesticide do I use? and 2) How do I apply it? Consequently, a whole generation of entomologists, pathologists, weed scientists, pest managers, and growers were trained in an approach to pest management that emphasized only one solution to pest problems—chemicals.

Interestingly, the idea that synthetic pesticides were the answer to all pest management problems was very short lived. Beginning in the 1950s, growers encountered problems with the use of these pesticides. Many pests started to show signs of resistance to the chemicals. Even though one individual of an insect species looks exactly like another, they are just as genetically diverse from each other as one human is from another. One outcome of this is that in a given population of an insect or disease species there is liable to be one or more individuals that have the genetic ability to detoxify a given pesticide and survive exposure to it. The susceptible individuals are killed off by the pesticide, leaving only the resistant ones to breed, resulting in a population of pests resistant to that pesticide. Growers found that as time went on it took more frequent spraying with heavier doses to control a specific pest. In some cases the pests became so resistant that they could no longer be successfully controlled by that particular pesticide. Furthermore, some pests became resistant to many different types of pesticides, a condition called cross resistance.

In other situations growers started to see pests that they had never encountered before. This phenomenon was termed a secondary pest outbreak. An insect or mite that was previously at very low population levels now occurred in epidemic numbers because a pesticide used against another pest species killed the agents (predators and/or parasites) that kept this "secondary pest" at low population levels.

Some pesticides turned out to be very toxic to wildlife such as birds. Rachael Carson's seminal book *Silent Spring* published in 1962 connected the death of many birds to the use of pesticides, instigating a backlash to their use. Moreover, toxic effects to humans began to surface.

DEVELOPMENT OF THE *IPM* APPROACH

The combination of problems that arose from overuse of pesticides, such as pesticide resistance, secondary pest outbreaks, environmental contamination, and effects on human health, led a forward-looking group of entomologists at the University of California to conclude that we were heading toward a pest management crisis in agriculture. They realized we had gotten away from the fact that pest problems are complex and ecological in nature. They concluded that the solutions to complex ecological problems must be broad-based and ecological in nature. These researchers developed the IPM concept to address the pest management crisis. Since its inception in 1959, IPM has evolved into the best way to manage pest problems on the farm (DeBach and Schlinger 1970; Huffaker 1980; Stern et al. 1959; etc.).

Integrated Pest Management (IPM) is an ecosystem-based strategy that focuses on long-term prevention of pests or their damage through a combination of techniques such as biological control, habitat manipulation, modification of cultural practices, and use of resistant varieties. Pesticides are used only after monitoring indicates they are needed according to established guidelines, and treatments are made with the goal of removing only the target organism. Pest control materials are selected and applied in a manner that minimizes risks to human health, beneficial and non-target organisms, and the environment.

Farming is an ecosystem-based and long-term endeavor, so we want to use management practices that are ecosystem-based and long-term in nature. By using a combination of control techniques to manage a pest problem, we develop a broad-based strategy that will still be successful even if one particular technique does not work. Also, based on our experience with chemical controls, we know that pest control decisions must take into account not only economic risks, but effects on the environment and people's health, as well.

Many definitions of IPM have been proposed since 1959. I like the following one best, in part because I can recite it without a cue card. IPM is a sustainable way to manage pests by combining biological, cultural, and chemical controls in a way that minimizes economic, environmental, and health risks. More importantly, I like this definition because it uses the words "economic, environmental and health," linking it to the definition of sustainable farming, which uses the words "economic, environmental and social." In my opinion, IPM is one of the cornerstones of sustainable farming. It is a systems-based approach to managing pests, just as sustainable farming is a systems-based approach to managing a farm. The IPM concept shares other attributes with the sustainable farming concept. IPM is a journey not a destination; it is a paradigm not a recipe. Like sustainable farming, I feel it is best viewed as a continuum from no IPM on the one hand to complete IPM on the other. And finally, because of these attributes, the concept of IPM, like sustainable farming, can be challenging to understand and implement.

There are five essential components to an IPM program (Ohmart et al. 2008; Figure 13-5):

1. *Understanding the ecology and dynamics of the crop.* It is important to gather all of the available knowledge about the crop one is growing. Many grape pest problems can be directly related to the condition of the crop and its interaction with the surrounding environment. The more one knows about the ecology of the crop, the better the pest management decisions will be. For example, it is well known that in California overly vigorous grapevine canopies can result in larger leafhopper populations than those observed on vines of less vigor.

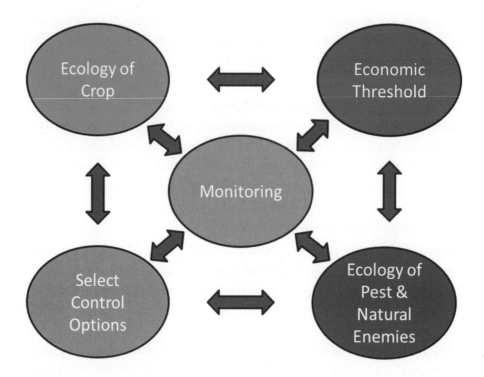

Figure 13-5 The five components of an IPM program.

Therefore, maintaining balanced vines is one way to keep leafhopper populations at acceptable levels (and to accomplish other goals, such as improved
wine quality). In the Eastern US, site selection is very important, encouraging
air flow in and around the vineyard, avoiding wooded areas that habor native
vines that are hosts for diseases and insect pests (Mark Chien, pers. comm.).

2. *Understanding the ecology and dynamics of the pest(s) and their natural enemies.*
    It is not only important to know what pests are present, but also to know the
    details of their life cycles, what makes their populations change, whether any
    natural controls are present, and what effects these may have on the pests. We
    may find some weakness that can be exploited if we know as much about the
    pest as possible.

3. *Instituting a monitoring program to assess levels of pests and their natural ene-
    mies.* It is vitally important to continually monitor the pest levels in the field.
    This is a crucial aspect of the IPM approach. By knowing how many pests are
    present, the best decision will be made regarding how much damage they might

cause to the crop. If natural enemies are present it is important to know how many are present, because they may take care of the pest problem for us.

4. *Establishing an economic threshold for each pest*. Effective monitoring and using economic thresholds make up the core of any IPM program in my opinion. What is an economic threshold? It is the level of a pest population above which, if a control action is not taken, the amount of damage caused by the pest will exceed the amount it costs to control that pest. In other words, it is the level of the pest population at which the control measure used pays for itself. It is important to note that factors such as paperwork time, interference with operations due to re-entry intervals, and possible secondary pest outbreaks need to be included in estimates of the cost of a pest management practice or program.

5. *Considering available control techniques and determining which are most appropriate*. A wide range of control techniques is available for crop pests. They can be divided into 5 broad categories: chemical controls, such as pesticides; cultural controls, such as controlling vine vigor or leaf removal; biological controls, such as natural enemy releases or conserving natural enemies; behavioral control, such as the use of insect pheromones; and genetic control, such as the use of resistant rootstocks or loose-clustered clones.

It is very important to choose the right control methods based on the economic nature of the pest problem, the cost of the particular control technique, and the effects of this method on the environment and people's health.

IPM Is an "Approach" and Changes with Time

IPM is not a technique or a recipe, but rather an approach to identifying, analyzing, and solving pest problems. Particular techniques for pest management may vary from field to field, year to year, crop to crop, and grower to grower, but the overall approach is always the same, using the five essential components of an IPM program. It is important to point out that an IPM program is not a cookbook approach. It would be nice if we could tackle a pest problem the same way every time, but history has shown us that this will not work.

Like sustainable winegrowing, an IPM program is never complete and is a process of continuous improvement. The reason for this is that over time we learn more about our crop, our pests and their natural enemies, and refine our monitoring programs. We also improve our economic thresholds, and develop new control strategies. We also periodically get new invasive pests. As we gain more knowledge, we need to use it to refine our IPM programs to make them more effective and to ensure they will work in the long term. This is the best way to minimize the economic impacts of pests in our vineyards and minimize the risks to our health and to the environment.

Some components of IPM are important enough to elaborate upon. The following paragraphs are devoted to them.

THE IMPORTANCE OF MONITORING IN SUSTAINABLE WINEGROWING

Vineyard pest monitoring is the foundation of any IPM program (Figure 13-5). Furthermore, monitoring in general is the foundation of a sustainable winegrowing program. Remember one of my favorite expressions "If you can't measure it, you can't manage it." Monitoring is the first step in measuring. Unfortunately, monitoring is often the most overlooked and deficient part of many growers' management programs. That is because monitoring takes time and time is money. Also, often it is challenging to develop a monitoring program that provides information that can be used in decision-making. This is particularly true for pests that are hard to monitor, such as cryptic pests like vine mealybug or mobile pests like Japanese beetle.

WHAT IS MONITORING?

This is a very simple question to answer: monitoring is keeping track of things (Figure 13-6). In vineyard management it is keeping track of what is happening in the vineyard. The first thing that comes to mind by many when hearing the word "monitor" is keeping tabs on pest populations. However, monitoring also involves keeping track of soil moisture and nutrient status, vine nutrition (via petiole samples), weather monitoring, measuring ET's, vine growth, and so forth.

Because I think vineyard monitoring is so important, and because it is an aspect of vineyard management that is often neglected, I continually try to think of analogies to illustrate its importance. One is to view monitoring in the same way as an instrument panel on a pickup truck. All the gauges on the panel keep the driver informed as to what is happening with the truck's various systems, which are very complicated and getting more so all the time. Without this instrument

Figure 13-6 Monitoring a vineyard for mites and leafhoppers.

panel, one would know almost nothing about the truck except when the engine was running because one would hear it, when it was moving, and when it ran out of gas because the truck would stop. The same can be said of our vineyard monitoring. It keeps us informed about the operation of our vineyard, which is an extremely complex biological system. With minimal monitoring (equivalent to the lack of an instrument panel) we are almost completely in the dark as to how our vineyard is operating, and as a result, we are not able to make the best decisions regarding its operation.

Continuing the analogy between the truck instrument panel and vineyard monitoring: as truck systems get more complicated, new gauges are added to the instrument panel to help drivers better assess how the truck is performing. Similarly, as we learn more about grapevines, soils, weather, and pests, we should be adding "gauges" to our monitoring program to help us manage our vineyards better.

WHY IS MONITORING IMPORTANT?

This is also a simple question to answer. Monitoring will help us better manage our vineyards. Another reason monitoring is important is that it helps ensure that only practices that are necessary will be implemented. One of the goals of IPM as of well as sustainable viticulture is to minimize vineyard inputs. Monitoring is the way this is achieved, since inputs will be added or actions will be taken only when our monitoring tells us it is necessary. One of the reasons pesticides were overused in the 1950s and 60s was because they were relatively inexpensive and at the time did not seem to have unintended consequences. In many cases they were applied as insurance and little or no monitoring was done to determine if applications were necessary. One of the breakthroughs of IPM was to demonstrate that monitoring is essential to minimize the use of pesticides.

A sound monitoring program has several important attributes.

Reliability

This may seem obvious but should not be taken for granted. A monitoring program needs to be so reliable that anyone who follows the procedures for that program will obtain similar results. Using equipment to make measurements, such as weather stations, helps ensure that results are the same regardless of the observer. Developing reliable methods for estimating the number of insect or mite pests or disease levels is much more difficult due to variation in distribution of pests within the vineyard. Quantification of a monitoring scheme helps a great deal in increasing the reliability of a monitoring program. For example, instead of reporting that a leafhopper population was light, moderate, or heavy, it is more beneficial to measure the population in nymphs per leaf.

Consistency

It is important that a monitoring program is carried out in the same manner following the same procedures each time it is done, so that the results are consistent from one monitoring to the next. For example, if measuring leafhopper populations one would not want to sample from only one area one time and then ten areas the next time. It would be better to take the same number of samples each time, ensuring they are representative of the overall vineyard.

Frequency

Monitoring should be done frequently enough to pick up trends or to spot developing problems. "Frequently enough" means you can watch the problem develop and still allow yourself enough lead time to consider management options. One of the best examples of this is monitoring mite populations. Say you visited your vineyard and found a significant mite population on the foliage. What you would do about this potential problem would probably differ considerably if you could not get back to check the vineyard for at least three weeks, as opposed to a situation in which you knew you would be back in a week. In the first scenario you would be much more likely to spray for mites than in the second one. By returning in a week you provide the opportunity for mite natural enemies to build up, for example, and yet still have time to treat the vineyard if necessary based on what is observed when revisiting the vineyard in a timely manner.

Speed

I like to ask growers and Pest Control Advisors (PCAs) the question, "What, besides money, do we never have enough of?" Invariably, the answer is "time." For a monitoring program to be useful to growers and PCAs it must be quick. However—and this is important—there is an inverse relationship between time spent monitoring and reliability. The more time spent monitoring, the more reliable the results will be. Therefore, the more efficient the monitoring, the more cost effective the program is overall.

Written Records

In my experience, this is most often the biggest weakness in a grower's or PCA's monitoring program, particularly in pest monitoring. Historically, very few people have kept written records of their pest monitoring. If you think I am wrong, ask your fellow growers or PCAs if they keep written records of their pest monitoring. I predict few will answer "yes," although I believe this is slowly changing. If you need further proof of the lack of pest monitoring, try to find a data management software package designed for farming operations that contains a decent component for recording and summarizing information on pest populations. A few years ago I saw a flyer for a computer software package that had 45 different components for handling every imaginable type of data, yet not one of them handled pest monitoring data. About 10 of these modules helped growers keep track of various types of data related to spraying for pests, such as spray recommendations, pesticide use reports, and worker safety. I found it amazing that there was no option for recording pest numbers growers or PCAs could use to help make decisions to spray the materials the software so nicely kept track of! The lack of such components is a good indication of the low priority pest monitoring occupies on the lists of many people in agriculture. The increasing use of handheld data logging devices or smart phones that sync with computer databases is making it easier to record pest monitoring in the field (Figure 13-7).

Why are written records of pest numbers so important? Because it is through correlating pest numbers and vineyard performance that we will be able to constantly improve our knowledge of economic thresholds, which is how best to minimize pest control procedures, whether they are chemical sprays or anything else. The only way we can accurately analyze the effectiveness of our pest management programs is to look at written records of pest monitoring. I am sure no one would consider trying to manage the financial aspects of their farming operation without keeping accurate financial records. Why should pest management be any different?

ECONOMIC THRESHOLDS FOR PEST MANAGEMENT IN VINEYARDS

An economic threshold is the level of a pest population above which the crop damage caused by the pest exceeds the cost of controlling the pest. In other words, it is the break-even point. If pest numbers are below the economic threshold and you treat for it, you are actually losing money because the damage caused by the pest, if you did nothing to control it, is less than the cost of the control measure. Using economic thresholds in pest management decision-making works well for pests for which you can monitor and then act on quickly if the number reaches or exceeds the threshold. Unfortunately, economic thresholds cannot be used in managing pests like plant pathogens, because once the pathogen infects a plant or its fruit it cannot be eradicated from the plant. Therefore, treatment for diseases must be prophylactic, in other words made before the pathogen becomes established on the crop, usually when weather conditions are conducive to disease development.

An economic threshold is a very simple concept, and it is the foundation of IPM, but it can be difficult to apply in the vineyard. It is difficult for many reasons, a few of which are listed here. First, an economic threshold can be affected by the

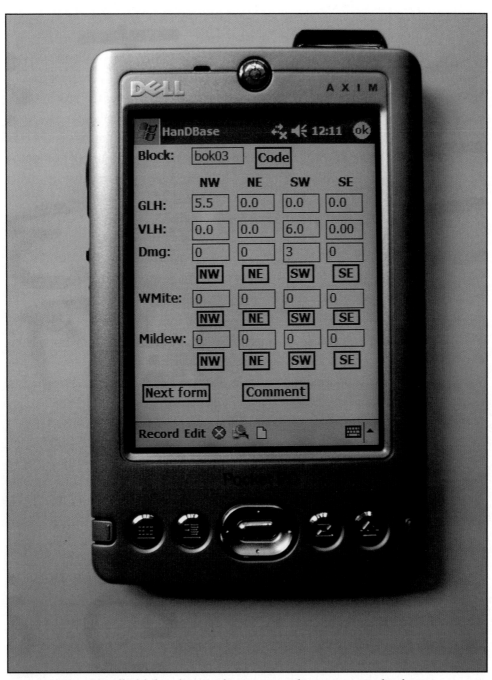

Figure 13-7   Handheld data logger that syncs with a computer database.

overall vigor of the vineyard. If vine vigor is low then it will take less pest damage to have a significant effect, and therefore the economic threshold is lower than in a more vigorous vineyard. Second, it can be affected by the time of year the pest damage occurs. For example, significant foliar damage just before harvest is usually less important than if it occurs at the beginning or the middle of the growing season, so an economic threshold for an insect or mite that causes foliar damage tends to get higher the closer you are to harvest. Third, it is dependent on the amount of pest damage that is already present. For example, a particular amount of leafhopper damage will be more significant if there is leafhopper damage present from earlier in the season. Trellis type can affect economic thresholds for leaf-damaging pests. On vines trained on a vertical shoot-positioned trellis there are fewer leaves shading the grape bunches when compared to vines on trellis types such as a "T" trellis or Geneva double curtain, so one cannot afford to have too much damage occur on these leaves for fear of losing them and exposing the fruit to too much direct sunlight. And finally, cost of the control measure affects an economic threshold. The higher the cost of the control measure the higher the economic threshold becomes.

Another major hurdle one confronts when attempting to use economic thresholds is that they have been scientifically determined for only a few vineyard pests. Why? For one thing economic thresholds are affected by many different variables, as discussed above, and it is very difficult and expensive to set up experiments that take them all into account when developing economic thresholds. Second, it is very difficult to correlate pest damage with crop loss, particularly for winegrapes, because so many variables seem to affect grape quality and we do not understand them in relation to pest damage. Third, particularly in more recent years, few research institutions are willing to carry out research on applied topics like developing economic thresholds for vineyard pests. In California, I am aware of economic thresholds for two vineyard pests; grape leafhopper and Willamette mite (Flaherty 1992). These thresholds were developed for Thompson seedless grapes and most people agree they are not very suitable for winegrapes. My experience as a Pest Control Advisor working in winegrapes, as well as in many orchard crops, is that most PCAs and growers have set their own thresholds for each pest based on experience. Interestingly, these thresholds can vary considerably from PCA to PCA and grower to grower. I will present an example that not only illustrates this point but also reinforces the importance of keeping written pest monitoring records.

From 1996 to 2006 staff of the Lodi Winegrape Commission monitored on a weekly basis important vineyard pests in 70 vineyards managed by 45 different growers and monitored by about 16 different PCAs. The monitoring was done on a quantitative basis and the data was input into a relational computer database. This was part of a larger Biologically Integrated Farming Systems (BIFS) project being carried out by LWC with the goal of increasing the implementation of sustainable farming practices in Lodi vineyards (Ohmart 2006).

Grape leafhopper was one of the important pests monitored during this project. Both the adults and nymphs feed on grape leaves, resulting in stippling where their mouth parts are inserted to feed, which in turn results in a loss of photosynthetic activity at these sites. It is not possible to monitor the adults accurately, but the nymphs can easily and quickly be counted and the results converted to a useful number of nymphs per leaf. If leafhopper nymphs and adults occur in high enough numbers, the feeding damage can be severe enough to delay fruit ripening, and if numbers are really high the feeding can result in defoliation, exposing the fruit to direct sunlight. If the defoliation occurs early enough before harvest the crop may not ripen at all. Sunburn may also occur along with dehydration of the fruit (Figure 13-8).

Figure 13-9 contains a graph in which the highest level of leafhopper counts, expressed as nymphs per leaf, are plotted for 29 vineyards that were not sprayed for leafhoppers and 27 vineyards that were sprayed for leafhoppers during the growing season in 2000. The sprays would have all been done during June and July. The leaf-hopper counts for both the sprayed and unsprayed vineyards were plotted from lowest on the left to highest on the right for each set of vineyards. Each point on the X axis of the graph contains a pair of vineyards, one sprayed and one unsprayed. The pairing and vineyard number are not significant, the position of where a vine-yard fell along the X axis is simply a result of where each vineyard fell out according to plotting them from low to high, left to right.

One of the first things to notice when looking at Figure 13-9 is that the data set of leafhopper counts for sprayed vineyards is very similar to the data set for leaf-hopper counts of unsprayed vineyards. In other words leafhopper counts found in sprayed vineyards ranged from low to high and in the same range of values as did the leafhopper counts in the unsprayed vineyards. This should not be the case if there was a consensus among growers and PCAs on an economic threshold. If there was such a consensus one would expect to see high counts, say about 10 nymphs per leaf or more, for all the sprayed vineyards and low counts for the unsprayed

Figure 13-8   Severe leafhopper damage that caused defoliation and fruit exposure to the sun.

vineyards. Instead, Figure 13-9 shows that some growers sprayed for leafhoppers when they were present in very low numbers while other growers did not spray for leafhoppers when they were present in very high numbers. As one would expect, the graph also shows that some vineyards were treated when leafhopper counts were high and others were not sprayed when counts were low.

What is Figure 13-9 telling us? When growers and PCAs decide to spray they do so because they feel the leafhopper populations are at a problem level or soon will be and the winegrape crop would be adversely affected. In other words, the numbers have exceeded a threshold that presents an unacceptable risk. The graph shows us that for the growers and PCAs involved in the project, this level varies from about two nymphs per leaf to 45 nymphs per leaf. Growers and PCAs who decide not to spray feel the leafhopper population present will not cause a problem. The range in leafhopper counts for this situation ranged from one nymph per leaf to almost 35. However, in the absence of monitoring data, this amazing range of

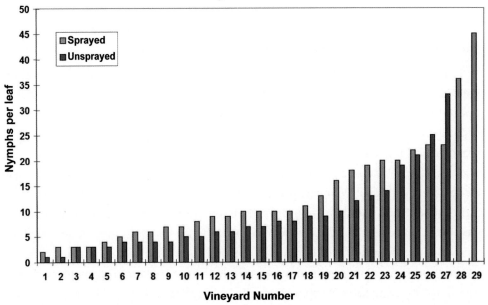

Highest Leafhopper Counts in Sprayed and Unsprayed Vineyards in 2000

Figure 13-9 Leafhopper counts expressed as nymphs per leaf in vineyards that were sprayed and others that were not sprayed in 2000.

values and that fact that it is similar for sprayed and unsprayed vineyards would not have been apparent. Without data one can only say "I thought there was a problem so I sprayed" or "I did not think there was a problem so I did nothing." Figure 13-9 shows us that the perception of risk varies significantly among growers and PCAs.

Since research has shown that grape canopies can suffer up to 20% defoliation without affecting yield, leafhopper counts of five nymphs per leaf or less indicate the population is not going to damage the crop. Therefore spraying when leafhopper counts are below five is not necessary. Likewise, growers and PCAs who are not treating when leafhopper counts are above 20 are likely suffering economic damage and should be spraying. Are growers and PCAs deliberately spraying when they do not need to and not spraying when they should? Of course they are not. The problem here is sorting out perceived risk of economic loss from real risk of economic loss. Collecting quantitative pest counts over time is going to help a grower figure this out.

The situation is made more difficult due to the fact that although research has shown grape canopies can suffer some leaf loss without the crop being adversely affected, an accurate economic threshold for grape leafhopper on winegrapes has not been established. So what can a grower or PCA do if there are no scientifically determined economic thresholds established for the pests they are managing and if research institutions are not going to work on them in the future? The answer is they need to determine them for their own vineyard operations.

I like to tell growers that one does not learn anything by spraying a pest population. That is because if the sprayer is correctly calibrated, is driven at the correct speed to get good coverage, and the proper pesticide is selected, the pest population will be controlled. This is a given. It is when one does not spray that something is learned, and it is one of two things. Either "Boy, I am glad I did not spray" because the pest did not cause any damage or "Gee, I wish I had sprayed" because damage occurred. If quantitative measures were made of the pest population when a spray is not made and if no damage occurs, then one knows the economic threshold has not been exceeded. If damage does occur when a spray is not made, one then knows the economic threshold was exceeded. In this trial and error manner, a grower can zero in on an economic threshold. I have spoken to many growers who recall not spraying for leafhoppers or mites and wishing they had because they suffered crop loss, whether due to yield or negative impacts on quality. However, since they did not record a quantitative measure of the pest and its damage, they had nothing to go on for the future except that they remember the pests were bad. "Bad" is not a useful measure when developing an economic threshold.

It is remarkable how many pest management decisions made in winegrapes, and many other crops, are based on perceptions rather than quantitative estimates of pest numbers vs. losses. It is understandable to some degree why this is the case. Growers and PCAs are very busy people and it takes time to make quantitative estimates. However, pesticides are expensive, as is the cost of their application, not to mention the "costs" associated with their environmental impacts, and every effort should be made to use cost/benefit analyses (i.e., economic thresholds) when making a decision whether or not to use them. As I have mentioned more than once, sustainable winegrowing is attention to detail. If you cannot measure it, you cannot manage it. Sustainable pest management decision-making needs to be data driven not perception driven.

I find it particularly interesting that cost/benefit analyses seem to be used even less often for environmentally friendly alternative strategies to synthetic pesticides. Just because something is "good" for the environment does not make it exempt from using the economic threshold concept. If the benefit of doing something does not exceed the cost to implement, it then it should not be done. If you are unsure of the cost/benefit ratio, then start trying to measure the costs and benefits so a sound decision can be made about its use in the future.

Growers are constantly talking about the "bottom line," yet cost accounting is not really driving pest management decision-making because we have no good economic thresholds. People are making these decisions based on their best guess. Applying the economic threshold concept is essential for successfully implementing truly sustainable vineyard operations.

The Three E's of Successful Spraying

Vineyard spraying is something that almost every grape grower will do multiple times during the year no matter whether they farm organically, biodynamically, or "conventionally." That is because most varieties of grapes are very susceptible to one or more diseases, no matter where they are grown in the US. Most regions also have their share of insect, mite, and weed pests that must be managed, often involving pesticide sprays. Vineyard spraying is an expensive operation because most spray materials and application equipment are expensive, and this equipment must be operated by people who are expensive to hire. Moreover, spraying can have environmental impacts due to the potential for off-site movement of spray material. Therefore one would expect growers would pay considerable attention to spray equipment and spray application. However, my experience is that the amount of attention focused on these issues is never what it should be. In part this is because growers are very busy people and must constantly cut corners to get all their work done.

If the decision to spray has been made, then it is important to approach it with attention to detail, just as one approaches all other aspects of sustainable winegrowing. This will ensure that one will achieve what Landers (2010) has termed the 3 E's of successful spraying: Effectively, Efficiently, and with due attention to the Environment. Landers emphasizes that the target is at the heart of effective spraying and needs to be clearly defined and the sprayer adjusted to provide the maximum coverage, optimizing the amount of material applied, and minimizing drift. Since pest control is the goal of most sprays, a successful spray is accomplished by what Landers calls four C's: applying the **C**orrect product to the **C**orrect target at the

Figure 13-10 Multiple row sprayer mounted on a self-propelled mechanical harvester. *Courtesy of Lodi Winegrape Commission.*

Correct time with the Correct machine. Although the correct machine is critical to a successful spray, like most aspects of agricultural production, successful spraying depends on good labor and good management. A motivated, well trained operator ensures that a spray will be successful and even though spray technology continues to advance, the need for good training continues (Landers 2010).

Effective vineyard spraying results from a combination of many things, the details of which are too extensive for coverage in this book. Landers (2010) does a great job covering these details in his book *Effective Vineyard Spraying*. However, there are a few facts that are important to point out here. Application rate per acre is determined by tractor and sprayer speed, nozzle size and pressure in the sprayer system. In general small droplet sizes are used for fungicide and insecticide/miticide sprays, moderate sized droplets for herbicides and large droplets for foliar fertilizers and pre-emergent herbicides applied to the soil. It is important to note that the slower the sprayer travels the more uniform the coverage, and the greater the speed the more variable is the spray deposition. Despite the critical importance of tractor speed, few tractors have accurate forward speed displays. In his book, Landers provides a detailed method for calculating tractor/sprayer speed and formulas for nozzle output.

If one is very concerned about reducing pesticide drift and minimizing the loss of pesticides to the ground one can use a tunnel sprayer, a type not used much in the West, which surrounds the vine canopy, capturing spray material that does not stick to the vine. Research trials have shown them to be the best spray drift reducing technology currently available, reducing drift by 90% compared to airblast sprayers. Furthermore, in early season sprays when vine canopy is minimal, tunnel sprayers recycle about 85% of the applied material. Tunnel sprayers have been shown to reduce the use of pesticides by about 35% when averaged over the entire season (Figure 13-11, Landers 2010).

Spray drift has become the most important issue for vineyard spraying, in part due to the increased urbanization of traditional rural areas where vineyards are located. There are two primary ways to address pesticide drift issues. The first is equipment related and the second is developing a spray drift management plan.

As mentioned above, the tunnel sprayer reduces drift more than any other option. If this equipment is not appropriate for your vineyard, consider using air induction nozzles. When properly used these can reduce drift by at least 50% (Ohmart and Storm 2008). The principle behind the nozzle is to create a larger spray droplet that will not drift but still maintain good target coverage. They are available as either flat fan or hollow cone, where an internal venturi creates negative

Figure 13-11 A tunnel sprayer.

pressure inside the nozzle body. Air is drawn into the nozzle through one or two holes in the nozzle, mixing with the spray liquid. The resulting droplets have tiny air bubbles in them, so the droplets explode on impact rather than bouncing off, creating coverage similar to that of much smaller droplets (Figure 13-12).

A spray or dusting drift management plan consists of a lot of common sense and can be broken down into five sections: identifying sensitive areas in and near the vineyard; being a good neighbor; establishing spray buffers; monitoring weather conditions; and, spray timing.

Dusting is included because it is a very common practice in many vineyards in California. While it is a very old method of disease management it is still the cheapest way to control powdery mildew in these vineyards.

IDENTIFY SENSITIVE AREAS

Identify any locations surrounding your vineyards where people, organisms, or structures could be exposed to spray or sulfur dust drift. These areas might include schools, bus stops, busy roadways, residences, other areas of human activity, nearby sensitive crops (pears), natural areas (vernal pools, riparian areas), or waterways.

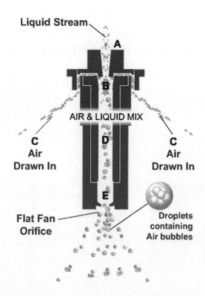

Figure 13-12 Diagram of an air induction nozzle.

BE A GOOD NEIGHBOR

Spray and sulfur dust stewardship includes being aware of the concerns of neighbors and local communities (Chadwick 2006). These actions enable mutual understandings and better relations, thus decreasing the probability of complaints. Below are some specific suggestions to consider:

1. Consider a policy of discussing vineyard actions with neighbors.
2. Form a regional team of growers to serve as the first contact with the public for negotiations and troubleshooting.
3. Take advantage of speaking opportunities with community organizations to explain the importance of sulfur for protecting crops against disease.
4. Periodically call your local county agricultural commissioner to check if spray drift incidents have been reported in your area.

It is important for growers to understand their legal rights related to farming, e.g., the right to farm regulations and local zoning designations and restrictions.

ESTABLISHING BUFFERS

Below are some ideas for establishing spray buffers:

1. Establish reasonable buffer zones to prevent drift onto sensitive areas and human exposure to applications.
2. Buffer distances vary with weather conditions, formulation (dust/wettable), application method (ground/air), presence of barriers (e.g., trees, hedgerows, open fields, etc.), and characteristics of sensitive areas.
3. If buffers determined for spray application overlap some border vine rows, apply separate fungicide sprays to these rows during conditions when buffers can be reduced.
4. Buffer Zone Procedures
   a. Leave an adequate buffer zone to protect sensitive areas.
   b. An adequate buffer zone is measured as enough distance for dust or spray to settle within the treatment area.
   c. Make sure applicator clearly understands the field being sprayed or dusted and the environment surrounding the field.

WEATHER CONDITIONS

Use the following to help determine what weather conditions you will use to decide when to postpone spraying or dusting:

1. Monitor weather conditions before and during applications.

2. Spray or dust should not be applied when the wind velocity exceeds 10 miles per hour, and consider using an even lower threshold.

3. Equip each sprayer/duster rig with an anemometer (wind-meter) to measure wind speed.

4. Avoid applications when winds are blowing towards sensitive areas.

5. Applicators should be aware that in some regions (e.g., Central Valley of California), "dead calm" conditions are often associated with a temperature inversion and applied sprays or dust can travel large distances. In these areas, applying sulfur or a spray when there is a minimum air movement of 2 miles per hour will help ensure that an inversion situation does not exist.

TIMING APPLICATIONS

The following techniques are suggestions for how to reduce drift problems with application timing;

1. Decrease public visibility and the potential for complaints by making applications during periods of least human activity (e.g., at night, weekends, etc.).

2. Develop a sequence for application that attracts less attention.

3. For nighttime applications, minimize "noise" complaints by treating rows closest to residential areas first.

4. Apply sulfur at times when minimum activity is occurring around your treatment area.

REFERENCES

Barbosa, P. and J. C. Schultz. 1987. *Insect Outbreaks*. Academic Press, Inc., New York. 578pp.

Chadwick, A. 2006. The Winegrape Guidebook For Establishing Good Neighbor And Community Relations. Calif. Assoc. Winegrape Growers, Sacramento, CA. 15pp.

DeBach, P., and E. I. Schlinger. 1970. *Biological Control of Insect Pests and Weeds.* Chapman and Hall, Ltd., London. 844pp.

Elliott, H. J., C. P. Ohmart, and F. R. Wylie. 1998. *Insect Pests of Australian Forests: Ecology and Management.* Inkata Press., Melbourne. 214pp.

Flaherty, D. L. ed. 1992. *Grape Pest Management.* Univ. Calif. Div. Agric. Nat. Res. Publ. 3343. 400pp.

Huffaker, C. B., ed. 1980. *New Technology of Pest Control.* John Wiley & Sons, New York. 500pp.

Landers, A. J. 2010. *Effective Vineyard Spraying: A Practical Guide for Growers.* Cornell Univ. Digital Print Services, Geneva, NY. 261 pp.

Ohmart, C. P. 1991. "The role of food quality in the population dynamics of chrysomelid beetles feeding on Eucalyptus spp." *Forest Ecol.* Mgt. 9:35–46.

Ohmart, C. P. 2006. *Ten Years of Lodi-Woodbridge Winegrape Commission's Biologically Integrated Farming System Program.* http://sarep.ucdavis.edu/BIFS/LWWCreport/.

Ohmart, C. P., and C. P. Storm. 2008. *Pest Management. In* Ohmart, C. P., C. P. Storm, and S. K. Matthiasson. *Lodi Winegrower's Workbook.* 2nd Edition. Lodi Winegrape Commission, Lodi CA. pp. 187–267.

Stern, V. M., R. van den Bosch, and K. S. Hagen. 1959. "The integrated control concept." *Hilgardia* 29:81–101.

14

Sustainable Viticulture

A significant amount of material has been presented in previous chapters that is a part of sustainable viticulture. One essential topic that has not been addressed yet is planting of the vineyard and management of the vines. In the context of this book I am going to call this viticulture, while realizing all that has been covered up until this point directly impacts managing the vines.

Many growers are starting to use the term winegrowing, rather than winegrape growing to describe what they do, because the final product that the customer buys is wine. For winegrape growers to be sustainable in the very competitive global market, the wine from those grapes must be in demand. It is well established that the customer's perception of wine quality is what drives purchasing decisions, regardless of the price point, and therefore the grower must have wine quality in mind from the time a decision is made to put in the vineyard through to when it is decided that it is time to take the vineyard out. I believe quality begins in the vineyard so it must be the goal of every winegrape grower.

I Am Not Going to present an exhaustive discussion of viticulture because the climactic conditions under which winegrapes are grown in the US vary tremendously, and how one manages the vineyards to maximize quality in each of these areas also varies greatly. There are many good print and web-based resources that the reader can seek out that describe good viticulture practices in various regions (Appendix *Resources*). I will point out some topics that are important in most regions that should be addressed in one's sustainable viticulture program.

VINEYARD ESTABLISHMENT

Vineyard establishment is the first and most critical step in premium wine grape production, regardless of the wine region in which the vineyard is planted. Optimum production efficiency, which is a key element in sustainable winegrowing, is achieved only through proper vineyard design and development. It is important to recognize that in most cases poor decisions made during vineyard establishment cannot be corrected until the vineyard is removed and replanted.[1]

Before establishing a vineyard, the grower should have a clear idea on specific production and fruit quality objectives, including the anticipated market tier for the grapes. This will help determine the proper vine spacing, training/trellising systems, and plant materials for the vineyard. These decisions cannot be made without thorough discussions with winery representatives. A winery contract should therefore be considered the first step in the vineyard establishment process. If the vines being planted are for your own use, be sure you are realistic about how much production you can use. It does not take many grapes to make a lot of wine and it is likely you will have grapes to sell. It is important to have a market for them.

Figure 14-1 A newly established vineyard in California.

---

1. Source: Storm et al. 2008.

The primary objective during vineyard establishment is to produce a healthy, well balanced vineyard—a vineyard in which fruit yield and canopy growth are naturally balanced and grape quality expectations are met. It is a widely held belief that the best winegrape vineyards in the world are those which require the least amount of manipulation during the growing season. Extensive canopy and crop load management practices are generally not necessary, at least in California, if the proper decisions were made during vineyard design and development.

The first step in planning the balanced vineyard is to understand the vigor or growth potential of the site. In addition to climate and annual rainfall, soil texture and depth are key factors when evaluating potential site vigor. Assuming no chemical or physical limitations in the soil, soil texture and depth combine to determine the water and nutrient reservoir available for vine growth. Simply stated, the larger the reservoir the greater the potential site vigor and vine growth. Understanding potential site vigor allows the grower to vary vineyard design parameters, including vine spacing, trellising and rootstock, to obtain the desired vine size and vigor level. It also will lead to fewer vineyard inputs, particularly with regard to irrigation and fertilization management, as well as canopy management practices such as shoot thinning, leaf removal, and hedging.

The vineyard production system is highly integrated. Manipulation of one parameter will likely have effects on several others. As mentioned in the Ecosystem Management chapter, the vineyard whole is greater than the sum of its parts. Therefore many factors must be collectively weighed when determining how vineyard design parameters will impact vine growth and productivity. Parameters known to have major effects on vine vigor and productivity, including rootstock, scion, plant spacing, and training/trellis system, must be considered collectively. It is important not to consider one of these parameters independently, without weighing its impact on the entire production system. For example, the correct trellis system for a given site may vary significantly based on the rootstock and plant density employed. Optimum vine spacing for the site will vary based on soil type, rootstock, and trellis system.

Storm et al. (2008) present in a self-assessment format the most important issues to consider when establishing a vineyard in the Lodi winegrowing region. Many of these are important to consider in other regions as well. I will briefly list them here and the reader can consult this publication for more details.

Climate and Soil Surveys, Water Quality and Availability

Climate and soil surveys are generally the first step in determining the suitability of a field for winegrape production. When planting a vineyard in an existing major growing region the climatic adaptability of grapes is generally well established. However, you should seek additional site specific information where available, particularly regarding spring frost potential and rainfall patterns. Other important site factors to consider, depending on the region, are soil drainage as well as air drainage. With a wine industry now active in every US state there are some excellent sources of information. The Cooperative Extension Service at your state's Land Grant University is a good place to start. Many regions have grower organizations that have formed to encourage exchange of information on winegrowing, so check to see if one is available to you.

Evaluating an area's climate and soil suitability is much more challenging in a new production region where little is known regarding the suitability of the climate for grape production. In this case a detailed historical climatic survey, fully assessing temperature, rainfall, frost potential and other major climatic events, is essential. It is important to assess the quality and availability of water for the vineyard. Poor water quality can reduce vine yields and fruit quality. Some elements, such as sodium, chloride, and boron are toxic to grapevines, so it is important to know the concentration of these elements in irrigation water prior to planting the vineyard. If problems are identified, often steps can be taken to help mitigate or reduce the potential detrimental effects of poor water quality on vine performance. Both scion and rootstocks should be carefully selected on the basis of their relative sensitivity to toxic elements if water quality is an issue.

Water quality also determines the suitability of water for use in drip irrigation. High levels of calcium bicarbonate or manganese, for example, may lead to clogging of the drip irrigation emitters. In addition, nitrogen levels in the ground water should be analyzed and taken into consideration when determining the fertilization requirements of the vineyard.

Environmental Survey

All forms of development, including vineyard development, impact natural resources and the environment. Vineyard development requires changes to the landscape that affect immediate and surrounding land, air, water, and living organisms. Vineyard development activities can include land clearing, tree removal, herbicide applications, riparian vegetation removal, brush burning, grading, disking, deep-ripping, re-contouring, altering water drainages, excavating, installing erosion control measures, and construction of roads, dams, wells, and fences. Many of these activities can affect natural resources and the environment. Regulatory programs exist to protect them. By examining your proposed or existing vineyard property and creating an environmental due diligence survey, you can help to safeguard against environmental damage and ensure that no laws are broken.

An environmental due diligence survey is an inventory of your property's physical characteristics that may affect farming the site and also may be subject to local, state, or federal regulations. These physical characteristics may include driveway and road systems, water access rights, streams and riparian corridors, vernal pools, wet swales, drainages, degrees of slope, existing erosion, and the presence of animal and plant species, such as oak trees or threatened or endangered species. These regulations vary considerably from region to region and county to county.

Begin your survey with a site description. It is very useful to include a map of the existing vineyard area or proposed vineyard development area with all of the following that apply: total acres, driveways, road systems, water access rights, streams and riparian corridors, vernal pools, wet swales, drainages, degree of slope, any existing erosion, and the presence of animal and plant species including any threatened or endangered species that may affect farming on the site and/or be subjected to local, state, or federal regulations. The program Google Earth, available for free on the Internet is an easy way to get an aerial view map of your potential vineyard site. You may also want to include photographs of your site and of sensitive areas on your property (e.g., wetlands, vernal pools, etc.). It is also a good idea to include information about your proximity to neighbors, schools and anything else that may be affected by your farming operation.

Record these sensitive areas on your site map and indicate on a separate document what measures you will use to mitigate adversely affecting these areas with the vineyard development and day-to-day management operations.

It is important to consult local, state, and federal laws that may pertain to your proposed vineyard site. For California growers, use the University of California publication titled *Growers Guide to Environmental Regulations & Vineyard Development* (2000). In other regions contact the appropriate agencies.

SOIL SAMPLING FOR PHYSICAL, CHEMICAL AND BIOLOGICAL CHARACTERISTICS

Soil evaluations are best performed by digging soil observation pits with a backhoe in several places on the property (Figure 14-2). These pits allow you or a soil scientist to collect samples for chemical analyses at various depths beneath surface, as well as to examine the soil profile and record notes on its texture, potential rooting depth, and the presence of impermeable layers. The number of pits necessary to perform the analysis varies depending upon the site. In sites with uniform soils, for example, no more than one observational pit for every 20 acres may be necessary. In contrast, sites with highly variable soils may require one observational pit for every four or five acres. In a replant situation, growers generally have a good idea where soil variability exists based on the growth uniformity of the previous vineyard. In this case observational pits may be more heavily focused in the areas with high variability.

Figure 14-2   Soil pit being examined at a potential new vineyard site.
Courtesy of Lodi Winegrape Commission.

A complete understanding of the physical aspects of the soil profile, including the presence of gravel or sand streaks, clay or hardpan layers, or abrupt changes in soil texture, is a critical component of proper vineyard establishment. This information is used to determine proper soil preparation steps prior to planting, including the use of deep ripping or slip plowing. Ripping is used to break or disrupt restrictive layers beneath the soil surface that may reduce effective rooting depth and drainage. Slip plowing is used to break restrictive layers as well as to mix the soil, effectively blending the soil near the surface with the soil below in order to create a more uniform profile. The correct depth and spacing of these practices is determined by examining the physical and chemical properties of the soil profile.

Variability within the soil profile is one of the key factors influencing vineyard uniformity. Unfortunately, relatively few steps can be taken following vineyard development to improve site uniformity. It is therefore critical to understand soil variability prior to planting and take the appropriate steps to improve uniformity. In addition to ripping and slip plowing, rootstock selection, plant spacing, block layout, and irrigation system design can be used to help reduce block variability.

Vineyard site evaluation is a very complicated process requiring a range of analytical skills that are not often possessed by the average grower. Fortunately, there are private consultants present in many regions who offer their services in helping analyze and select the right vineyard site based on a grower's goals and needs.

TAKING NEMATODE SAMPLES

Nematodes are microscopic worms. Of the many species, most are beneficial and eat decaying organic matter or organisms such as bacteria, fungi, and other nematodes, but some eat plant roots. These are called plant parasitic nematodes. Each species of plant parasitic nematode differs in its feeding habits and how it affects the various rootstocks, so samples must be taken before planting a vineyard to make the correct nematode resistant rootstock decision.

Before sampling, if there are differences in soil texture, moisture, drainage, plant health, and cropping history the site should be subdivided. For each subdivision take a separate sample.

Samples should be taken when the soil is moist. Include healthy roots of the previous perennial crop if possible. It is important to take samples down to the 3-foot level. Take at least 15–20 sub-samples every five acres and mix together (Figure 14-3).

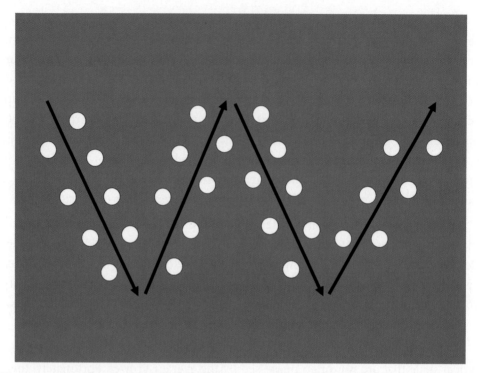

Figure 14-3 This is an example of how to sample for nematodes in an open field before planting (from Storm et. al. 2008). *Courtesy of Lodi Winegrape Commission.*

Remove about one quart from the mixture, fill a plastic bag and place it in an ice chest (ideal temperature is 40° to 50°F—not too cold, not too warm). Mail the samples to the lab as soon as you can.

Frequently, growers choose to take fewer samples from larger areas, typically one sample for every 20 acres. If this is economically necessary, then increasing the number of sub-samples per sample can help to offset the loss of accuracy from this more general sampling.

ROW ORIENTATION

In warm regions optimum fruit quality is obtained when clusters are exposed to adequate amounts of dappled or indirect sunlight—for example sunlight reaching the fruit zone as sunflecks through small openings in the canopy surface, or sunlight filtered through a single leaf layer. Smart and Robinson (1991) brought this approach to the attention of the winegrowing world. Direct sunlight per se is not detrimental to fruit quality, but the related increase in fruit temperature as a result of exposure to full sunlight can cause sunburn on the fruit and cooked flavors in the wine.

For east/west orientated vineyard rows, the southerly facing fruit is exposed to extremely intense heat and sunlight. Conversely, northerly facing fruit is shaded for most of the day.

With north/south oriented vine rows, the westerly facing fruit receives the hot afternoon sunlight and the easterly facing fruit receives an equal amount of sunlight, but it is the cool morning sunlight. These differences between exposed and shaded fruit can cause significant harvest and flavor differences.

By aligning vineyard rows in the NE/SW orientation, the south-easterly facing fruit is exposed to a longer period of cooler morning and early afternoon sunlight while the north-westerly facing fruit still receives adequate sunlight hours, but is exposed to the intense afternoon heat for a much shorter period (Figure 14-4). The disadvantage to this system in California is that most property boundaries are aligned on a NS and EW axis so you will end up with some very short rows and some very long rows because of it, making farming difficult and likely increasing the cost of vineyard trellising.

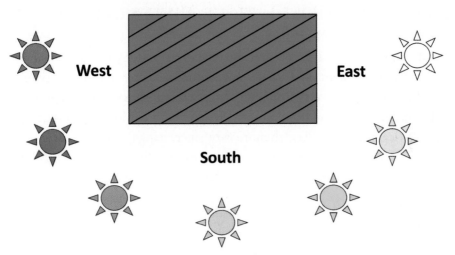

Figure 14-4 Schematic showing the advantages of a NE/SW row orientation in a climate like the warmer regions of California (from Storm et al. 2008). *Courtesy of Lodi Winegrape Commission.*

VINE SPACING AND TRELLISING

Plant spacing is one of the most important decisions when planting a new vineyard. Row spacing, or the distance between vineyard rows, determines the amount of fruiting area or the number of linear feet of fruiting surface per acre. Recent work in California has shown that for single curtain canopies, such as the California Sprawl, an 11% increase in yield potential is obtained for every one foot reduction in row spacing (i.e., as row spacing is decreased from 11 feet per row to 10 feet). In practice, vine rows should be placed as close together as possible in order to maximize yield potential while maintaining ease and practicality of farming practices.

In-row spacing or the distance between vines within the row is largely determined by anticipated vine vigor. The optimum in-row spacing allows vines to be trained completely (the cordon or fruiting wire fully filled) entering year three or their first cropping year. Vines spaced too far apart within the row may not be fully trained by the end of the second growing season, requiring more time to reach full production. Vines spaced too close together can be too vigorous, resulting in excessive shading and the need for more extensive canopy management practices, including shoot thinning and basal leaf removal. The cost of establishing and farming these vineyards is also greater due to the increased number of vines per acre.

STAKE AND END POST FATIGUE, VINEYARD DESIGN AND HARVESTING METHODS

When designing a vineyard, consider the type of harvester you will be using if you intend to use a mechanical harvester. A bow rod harvester that shakes the canopy affects the trellis system very differently from a trunk shaking harvester. Consider also the cordon height and training style. Systems with cordons high off the ground or with vertical training will catch more wind, increasing the horizontal leverage on the stakes, fatiguing the metal or wood and eventually bending or breaking the stakes. For these systems, it is recommended to use thicker gauge stakes and end posts.

ROOTSTOCKS AND CLONES

In viticulture, rootstocks are used primarily to provide resistance to soil pests such as phylloxera and nematodes. However, rootstocks provide many other horticultural benefits. Rootstocks can significantly influence vine vigor and can be used to increase or reduce anticipated vine size depending upon soil type, growing conditions, and scion variety. They also provide resistance to drought, as well as conditions of persistent soil moisture or high water tables. Rootstocks can also be used to help vines perform better under high soil salinity or serpentine conditions, as well as in situations where the pH is low or high. In addition to their pest resistance properties, the many other characteristics of rootstocks should be given careful attention when selecting the best rootstock for a new planting.

Growers should pay careful attention to the viticultural traits of specific clones when selecting plant materials for a new vineyard. Basic fruit morphological traits, such as cluster compactness and berry size, have dramatic effects on bunch rot susceptibility. These characteristics, along with vine productivity and growth habit, should be given particular consideration. Data from the clonal evaluation trials performed by University of California Farm Advisors and other researchers are particularly useful for this purpose. If this information is not readily available for a particular clone you are interested in, contact your nursery or winery to see if other growers have planted the clone and ask if you can view the vineyards prior to making your decision.

Scion and rootstock material should be certified virus-tested, or at least tested by a reputable lab. Viruses can be latent, which means that the wood can be infected, but the rootstock or the site is not correct for the symptoms to appear. However, when the wood is grafted or budded into your vineyard, on a different rootstock, surprise... the virus shows up. This is why certification or testing is highly recommended.

VINEYARD NUTRITION MANAGEMENT

Most winegrape growers have a nutrition management plan for their vineyard. However, the form in which the plan exists and the level of detail in the plan likely varies significantly from one grower to another. For example, for some the plan exists primarily as a mental record while for others it is written down. The benefits of keeping written records of farming operations have been highlighted in previous chapters, and creating a written vineyard nutrition management plan is no exception.

There are many reasons to develop and implement a vineyard nutrition management plan. The most important is to optimize the amount of nutrients that are applied to the vineyard. This will likely save you money because only the amount required to achieve your yield and quality goals will be applied. Furthermore, optimizing nitrogen applications not only will help your bottom line, but it will minimize greenhouse gas production (GHG) occurring in your vineyard. Nitrous oxide has about 320 times more greenhouse gas potential than CO_2, and fertilization of your vineyard, either with natural fertilizers such as compost or manure, or with synthetic nitrogen fertilizers, results in its release into the atmosphere due to soil microbial activity on the fertilizer or compost. Having vineyard nutrition carefully planned and managed means only the amount of nitrogen that is absolutely necessary will be added, minimizing GHG production.

One of the goals of organic or biodynamic vineyard management is to reduce the amount of off-farm nutrient inputs by growing cover crops, adding compost made on the farm, manures from on-farm livestock, and encouraging soil microbial populations. However, it is important to realize that nutrients leave the vineyard in the grape crop. So no matter how biologically active the vineyard soil is, if enough grapes are taken out of the vineyard a nutrient deficiency will develop at some point in time. Some nitrogen is fixed from the atmosphere by leguminous plants like vetch and fava bean, so growing them will ensure that nitrogen is added to the soil. However, grapevines also require many other nutrients, so when soil nutrients become depleted after multiple harvests they will need to be replaced by nutrient inputs from outside the vineyard.

A good way to visualize a vineyard nutrition management plan is to think of it as a budget. Nutrients leave the vineyard in the harvested grapes. To keep them balanced in the vineyard soil an equal amount needs to be returned, assuming none were limited to start with. What nutrients and how much of each leave the vineyard must therefore be determined so one knows how much to return. The kinds and quantity of nutrients leaving the vineyard can be readily determined by analyzing the fruit and yield per acre. The challenge is in figuring out what form and how much of each nutrient should be added to the vineyard to maintain nutrient balance in the soil.

The availability of nutrients in the soil to vines is affected by many factors. First, vines take up nutrients at varying rates during the growing season based on the growth stages of the canes, the leaves, the fruit, and the roots themselves. Some nutrients like nitrogen are soluble in water and if not taken up by plants can either run off the vineyard in surface water or leach out of the soil into the water table. Some nutrients like potassium bind to soil particles to varying degrees depending on the soil type and can become unavailable to the vine. Rootstocks vary in their ability to take up nutrients. And finally, if one uses compost for vineyard fertilization, it is difficult to determine the amount of available nutrients at any given time, because they become available only through breakdown of the compost by soil microbes. The rate of decomposition will vary with soil temperature and moisture.

AN EXAMPLE VINEYARD NUTRIENT BUDGET

The easiest side of the nutrient budget equation to calculate is the kinds and amount of nutrients leaving the vineyard in the crop. That is because reference data exists for this information. An example is presented in Table 14-1.

Now it is time to tackle the more difficult side of the equation, which is to determine what vineyard inputs are required to balance the nutrient budget. First list all of the ways nutrients can be put back in the vineyard. The various possibilities are from leguminous cover crops, irrigation water, and inputs such as compost, manure, or synthetic fertilizers. Then determine to what extent you are willing to go to estimate the nutrient contribution of each one of these inputs. Some are much easier than others to calculate.

Determining the nutrient content of irrigation water is done with a simple water test. It is important to realize that irrigation water from some sources contains nitrogen and can end up adding a significant amount over a season of irrigation. For example 10ppm of nitrogen in irrigation water results in 27 lbs of nitrogen in an acre foot of water.

Another easy calculation is the amount of nutrients being added in synthetic fertilizer. The label on the container should have the numbers you need. However, keep in mind that adding 10 lbs of nitrogen or 20 lbs of potassium per acre does not mean an acre of vines will take up that amount. That is because soil type and soil moisture availability affects the ability of a vine to take up the nutrients, as does its rootstock and, as mentioned above, vine nutrient demand varies depending on growth stage of the plant.

If natural soil amendments are applied, such as compost or manures, it is important to determine the amount of the major nutrients being added to the vineyard.

Table 14-1 The amounts of major nutrients removed from the vineyard in harvested grapes (Storm et al. 2008).

Tons/acre of grapes	N	P	K	Ca	Mg
1.0	2.9	0.6	4.9	1.0	0.2
1.5	4.4	0.8	7.4	1.5	0.3
2.0	5.8	1.1	9.9	2.0	0.4
2.5	7.3	1.4	12.4	2.5	0.5
3.0	8.8	1.7	14.8	3.0	0.6
3.5	10.2	2.0	17.3	3.5	0.7
4.0	11.7	2.2	19.8	4.0	0.8
4.5	13.1	2.2	22.2	4.5	0.9
5.0	14.6	2.8	24.7	5.0	1.0
5.5	16.1	3.1	27.2	5.5	1.1
6.0	17.5	3.4	29.6	6.0	1.2
6.5	19.0	3.7	32.0	6.5	1.3
7.0	20.5	4.0	34.4	7.0	1.4
7.5	21.9	4.3	36.8	7.5	1.5
8.0	23.4	4.6	39.2	8.0	1.6
8.5	24.9	4.9	41.6	8.5	1.7
9.0	26.4	5.2	44.0	9.0	1.8
9.5	27.8	5.5	46.4	9.5	1.9
10.0	29.3	5.8	48.8	10.0	2.0

Even though a nutrient amendment is "natural," one still needs to calculate what is being applied. Too little or too much of something will not allow you to achieve your quality and yield goals. If the compost is purchased, ask for a nutrient analysis to determine the amount and type of nutrient input per ton of compost. If manure is added to the vineyard an analysis should be done if one wants to establish the exact amount of nutrients being added by the manure.

Figure 14-5   Forage cover crop with high biomass per acre that on a low vigor site would compete with vines for water and nutrients.
Courtesy of Lodi Winegrape Commission.

If a leguminous cover crop is planted in the vineyard, it is important to calculate how much nitrogen it is adding per acre. It is possible for a good nitrogen fixing cover crop to contribute more than 60 lbs per acre in a season. The simplest way to make this calculation is to take a 3 ft by 3 ft sample of the cover crop and send a 1 lb subsample of it to a chemical lab for a moisture content and dry weigh nitrogen content analysis. The results can be used to calculate how much nitrogen per acre the cover crop is contributing. Unfortunately not all of the nitrogen from the cover crop is likely to be taken up by the vines (Berry et al. 2001). While on the topic of cover crops, do not forget that some can be a nutrient drain on the vineyard, competing with the vines for nutrients. This can be used to your advantage in some situations and can cause problems in others.

Once you have determined the types and amounts of nutrients that are added to the vineyard by the inputs in your vineyard nutrition program, it is time to tackle the most challenging part of the nutrient management plan, which is to take into account the factors that affect nutrient availability in the vineyard and vine nutrient demand over time. In general, factors that affect nutrient availability will influence the total amount of nutrients that must be applied during the year, whereas factors that affect vine nutrient demand will dictate the timing of nutrient additions throughout the year. If one primarily uses compost for vineyard nutrition, the timing of application of the compost is not nearly as critical as it is for water-soluble synthetic fertilizers.

A detailed soil map of the vineyard is the best way to determine the spatial variability of nutrient availability to the vines. Hopefully, this was done when the vineyard was planted, so that the root stock matches the soil characteristics in terms of achieving the nutrient uptake desired for your quality and yield goals. Looking at the variability of yield over the vineyard will also provide clues as to the variation in nutrient uptake over the site. Taking a soil sample every three years or so will help identify nutrient imbalances, pH problems (which can affect nutrient availability), and track cation exchange capacity (CEC), so that problems can be addressed and the performance of corrective actions tracked.

Many viticulturists view the vine as the best indicator of vineyard nutrient status and use tissue sample analyses to help gauge the nutrient balance in the vine and the need for any nutrient additions. As with many viticulture techniques, tissue sampling methods vary significantly among growers. Some are happy with one petiole sample at bloom taken from reference vines while others take multiple tissue samples throughout the growing season. Many of the nutrient critical values for *vinifera* varieties were established by research more than 20 years ago, and some growers feel they are no longer relevant. There is also a debate over whether petiole samples or leaf blade samples are the best indicators of vine nutrient status. I will not venture an opinion on what is best but will say that the debate indicates there are still significant research questions to be answered about the best way to assess vine nutrient status using tissue samples.

It is clear from the discussion in the last few paragraphs that deciding the amount of vineyard nutrient inputs to apply to keep the nutrient budget balanced is complicated. It will likely take a number of years to fine tune the amounts applied and adjustments will have to be made from year to year based on how well you are achieving your quality and yield goals, as well as variability caused by weather and climate.

Once one has gone through the above steps and gathered the information to the level of detail that meets one's vineyard nutrition goals it is time to put the plan down on paper. There are various ways to organize the information. One way is to create a small binder starting with a vineyard soils map, followed by the results of any soil sample analyses, water analyses, and vine tissue samples. Then list a historical record of the vineyard's yield and quality, along with important notes on weather for the year, crop maturity, bunch counts, and vine vigor. Follow this with a description of the forms of nutrients that will be added and their nutrient content. End with a detailed description of the types and amounts of nutrients to be applied, the timing of these applications, and application methods. It is also important to list any sensitive areas in and around the vineyard that could be affected by nutrient additions. Finally, write into the plan that you will review it annually and update it as necessary.

The International Plant Nutrition Institute has created an easy to remember model to follow for developing a sustainable nutrition management plan similar in structure to the one I described above. It is called the 4 R's which are: obtaining your plant nutrients from the **Right** source for your vineyard; applying the nutrients at the **Right** rate; ensuring the applications are being made using the **Right** timing; and making sure they are applied at the **Right** place in the vineyard (Figure 14-6). The IPNI website has some very useful articles describing this model that should be very helpful in developing your vineyard nutrition management plan.

The importance of a written nutrient management plan cannot be overemphasized. Few of you would consider managing your farm's financial budget totally in your head, so why would you manage your nutrient budget that way? Proper nutrient management is critical for producing quality winegrapes, and nutrient inputs are expensive. Proper nutrient management will also help you minimize off-site movement of nutrients, minimizing impacts on survey and ground water.

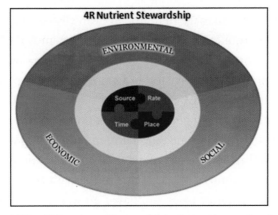

Figure 14-6   The 4 R's of Nutrient Stewardship http://www.ipni.net/4r.

VINE BALANCE

Consumer demand for higher quality wines has driven substantial changes in viticultural practices during the past two decades. If a grower cannot produce the right quality grapes for this wine, they will soon go out of business and therefore not be sustainable. While optimum practices vary due to growing conditions, cultivar and designated market tier, all premium winegrape growers should strive to achieve balanced vines. These are vines that have a natural balance between their vegetative growth or canopy size and fruit yield. Balanced vines have sufficient leaf surface to ripen their crop to optimum quality, without suffering a delay in harvest. The relationship between vine leaf area and fruit yield is defined as crop load or the amount of fruit produced per unit of leaf surface. Growers can estimate vine balance by measuring pruning weights during the winter and making a few simple calculations. It is important to realize that both high- and low-yielding vineyards can be well balanced, as long as their crop loads are within the optimum range. Some feel that only low-yielding vines produce the best quality grapes and wines. However, there is much research supporting the view that a balanced vine is what produces the best quality grapes and wine.

Dormant pruning is the most important annual cultural practice in wine grape production. It is the first step in determining vine cropping potential for the upcoming growing season, and also has a major impact on canopy size and growth characteristics. Pruning establishes the number of buds retained per vine, and therefore the number and size of fruit clusters at harvest. Pruning levels also determine shoot number per vine, impacting final shoot length and leaf size and number.

Canopy and crop load management practices are also important components of premium winegrape production. Shoot thinning, the removal of unwanted shoots in the early spring, along with basal leaf removal following fruit set, are effective practices for reducing canopy density and improving the microclimate and air movement within in the canopy interior when needed. Canopy microclimate—the temperature, humidity and sunlight within the canopy interior and fruit zone—is a critical factor in maintaining sustainable yields and producing high quality fruit. As mentioned earlier in this chapter, large amounts of indirect or dappled sunlight and modest levels of direct sunlight are desirable in the fruit zone. This allows fruit to ripen with the highest possible color and flavor. In warm climates clusters exposed to direct sunlight often become too hot, particularly in the mid- to late afternoon, causing the fruit to be lower in both color and flavor compared to fruit with moderate exposure to sunlight. Air movement within the canopy interior helps reduce humidity and potential disease pressure.

The specifics of how to achieve vine balance vary among regions and I will not attempt to review this material here. Refer to the resources available to you through University Cooperative Extension.

REFERENCES

Berry, A., and R. Smith. 2001. *Uptake of Cover-Crop Nitrogen by Wine Grapevines in the Central Valley of California.* Report to the Lodi-Woodbridge Winegrape Commission, Lodi CA. 9pp.

Horwath, W., C. P. Ohmart and C. P. Storm. 2008. "Soil Management." *In* Ohmart C. P., C. P. Storm, and S. K. Matthiasson. 2008. *The Lodi Winegrower's Workbook.* Lodi Winegrape Commission, Lodi CA. pp. 111–141.

Storm, C. P., C. P. Ohmart and N. Dokoozlian. 2008. "Vineyard Establishment." *In* Ohmart C. P., C. P. Storm, and S. K. Matthiasson. 2008. *The Lodi Winegrower's Workbook.* Lodi Winegrape Commission, Lodi CA. pp. 64–94.

15

The Role of Certification
in Sustainable Winegrowing

Ending this book with a chapter on sustainable winegrowing certification may see a bit unusual to the reader. However, I have good reason to do so. Over the years when I have made presentations to winegrape growers, particularly to those not familiar with what sustainable winegrowing means and what it entails, they have expressed the desire to become involved in sustainable winegrowing by developing a sustainable winegrowing certification program. While sustainable winegrowing certification is a laudable endeavor and can achieve important objectives, I do not feel it is a good way for a grower to be introduced to the concepts and practices of sustainable winegrowing. That is because certification means that some growers are "in" because they can qualify while other growers are "out" because they cannot. Furthermore, I think it is a real mistake for a grower group to become involved in sustainable winegrowing for the first time by designing a sustainable winegrowing certification program. Invariably during the process of developing standards, disagreements will arise among the group as to what is a good standard and what is not. The result is a divided group instead of one united in an effort to move along the sustainability continuum. Certification involves drawing a line on the sustainability continuum such that those to the right of the line are "in" and therefore by implication sustainable and those to the left are "out" and thereby labeled as not sustainable (Figure 15-1). I am not implying certification is a bad thing, but rather that if one wants to join a program or be a part of developing one, it is critical to know what one wants to accomplish by, doing so and to understand what certification is and what are the various kinds of certification. One should join a certification scheme with one's "eyes wide open."

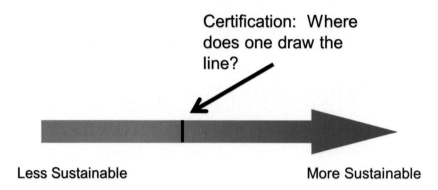

Less Sustainable   More Sustainable

Figure 15-1 Certification involves drawing a line on the sustainability continuum to indicate the minimum level of sustainability (© Dale Goff).

WHY CERTIFICATION?

Certification programs are created to add validity and credibility to a claim made about a product. Consumers do not necessarily believe a claim or story about a product. To address this lack of trust and confidence, many groups such as manufacturers, farmers, architects, as well as government and nongovernment organizations have created certification programs to reassure the consumer that product claims are valid and credible. I will restrict my discussion to sustainable certification programs for producing food and wine, but there are many certification programs for manufactured goods (e.g., energy star) and buildings (e.g., LEED).

A certification program for the sustainable production of food or wine consists of a set of standards that must be adhered to in order for them to be declared certified. An auditor inspects the process by which the product is processed or grown to verify the producer is meeting those standards. There are several ways a certification program can be created and I will list them in an increasing order of credibility. There is self-certification, where a group creates their own standards and does their own auditing to ensure the standards are being met. There is independent certification, where a group creates standards and has an independent auditor verify that the standards are being met. And finally there is third-party certification, where a group uses standards accredited by an independent or "third" party and the auditing is done by an independent auditor.

Assessing the credibility of a certification program is further complicated by the different ways a group can develop the standards. For example, a group can create their own standards without any peer review. Another method is for a group to form a committee of stakeholders to create the standards and then get them peer reviewed. A third is to create a set of standards through a stakeholder process and have them accredited by a third party. And finally a group could use a set of standards that was developed by a third party.

One of the many interesting discussions associated with certification focuses on what type of certifier is most trustworthy. Some feel that the government, whether local, state, or federal, is the most credible certifier. However, others do not trust the government and therefore feel that nongovernment organizations make the most trustworthy certifier. One thing on which many will agree is that self-certification is the least credible type of certification program and third- party certification is the most. Furthermore, a credible program should be completely transparent, meaning one should be able to easily find out what the standards are, what process was used to create them, whether they have been accredited and/or peer reviewed, and how the auditing process is carried out. The Internet makes transparency a much easier goal to achieve. A website for a certification program is a great place to post standards that are downloadable and read by anyone who wishes to do so.

If one is serious about using certification as a guide to making purchases, or is contemplating participating in one as a winery or grower, it is wise to carefully check out the program of choice to ensure it meets one's needs and is at an acceptable level of credibility.

TYPES OF CERTIFICATION STANDARDS

One very important aspect to understand about a certification program is the nature or type of standards on which it is based. There are three basic types: process-based, practice-based, and performance-based standards.

A process-based certification program is one where the participant is practicing a cycle of continuous improvement in their winery and/or vineyard. The California Sustainable Winegrowing Alliance's (CSWA) Sustainable Winegrowing Certification is an example.[1] Another is the international ISO 14000 program.[2] For those growers and winemakers who have participated in programs similar to CSWA's

---

1. *www.sustainablewinegrowing.org/swpcertification.php*
2. *www.iso.org/iso/iso_14000_essentials*

Code of Sustainable Winegrowing Workbook program, the standards one must meet in a process-based certification program will sound familiar. One must first assess their operation, most often using a self-assessment tool, such as the workbook mentioned above. Then one or more areas of concern are identified and an action plan is created to improve upon the concern. The action plan is carried out and after a specified time period, a year, for example, one does the self-assessment again to see if improvement has been made. If it has, then new concerns are identified and new action plans are created. If improvements have not been made on the area of concern identified in the assessment, the original action plan is revisited and altered if necessary to ensure improvement is made. This cycle is continued indefinitely and one increases one's level of sustainable practices over time.

In a process-based certification program the auditor verifies that one is doing the self-assessment, creating action plans, carrying out the action plans, and then re-assessing themselves over a specified time period. A very important thing to note about a process-based certification compared to the other two types is that one does not need to be at a specified level of sustainability to qualify for certification. Put in terms of a sustainability continuum, from not sustainable on one end, to completely sustainable on the other, participants can be at various points along that continuum and as long as they are participating in the cycle of continuous improvement they all qualify for certification. One is certified as participating in the cycle of continuous improvement, not as being at a particular level of sustainability.

Practice-based certification, which is by far the most common type, is one where the standards are practices that must be used and which are assumed to improve one's level of sustainability. Organic and Biodynamic farming are practice-based certification programs, as are the newer sustainable winegrowing programs such as the Lodi Rules for Sustainable Winegrowing,[3] Napa Green,[4] Fish Friendly Farming,[5] Oregon LIVE,[6] and Sustainability in Practice.[7] The auditor of these types of programs visits the farm or winery to ensure that the required practices are being carried out. In all practice-based programs of which I am aware, the required practices imply a specified level of sustainability as determined by the certifying body,

---

3. www.lodirules.com
4. www.napavintners.com/wineries/napa_green_wineries.asp
5. www.fishfriendlyfarming.org
6. www.liveinc.org
7. www.sipthegoodlife.org

Figure 15-2 Logo for the Stewardship Index for Specialty Crops program.

which in the case of organic farming is the USDA; for Biodynamic farming it is Demeter, and for Lodi Rules it is Protected Harvest.

The third type of certification program is one that is performance-based. In these programs the standards are specified levels of performance as measured by a set of metrics rather than a set of required practices. The metrics can be the amount of water used to produce the product, for example, or the amount of energy used, and so forth. At this point in time, I am not aware of any certification programs for agriculture that are performance-based. However, the Stewardship Index for Specialty Crops is in the process of agreeing upon a set of metrics that can be used to measure sustainability performance not only on the farm but all along the food supply chain to the grocery store.[8] If this project is successful, then it is likely some organizations will create certification programs with performance-based standards requiring the participant meet a specified level of one or more metrics that the certifier judges to be sustainable.

Each type of certification program, whether process-based, practice-based, or performance-based, has its strengths and weaknesses. Process-based programs are ones that most if not all can participate in, and ensure that everyone is improving over time. However, it does not require a specified level of sustainability be achieved. Practice-based programs require specific practices be implemented and most are quite rigorous. However, because they specify the practices that must be used, some find this constraining. Moreover, as research and experience are gained over time it is likely new practices should be added to the certification programs, but often this is difficult or impossible to accomplish. The advantage of a performance-based system is that although one has to achieve a specified level of performance, such as not exceeding a threshold of energy or water use, the program leaves it up to the participant to figure out what practices to implement in order to

---

8. *www.stewardshipindex.org*

meet that level of performance. Another advantage of performance-based programs is that they are based on outcomes. We assume that practice-based programs result in improvement. For example, we think growing cover crops, reducing tillage, and adding compost improves soil quality, but in a practice-based program soil quality is not measured. In a performance-based program the outcome, soil quality, is what is measured, and a specific level is required to qualify for certification. However, if no incremental improvement in performance is required over time, a performance-based program will be static like most of the practice-based programs.

Certification, like many things, has advantages and disadvantages. If one wants to make a value- added claim about grapes or wine, one way to add credibility to this claim is by participating in a sustainable certification program. However, before picking the right program I suggest sitting down and figuring out what you want the certification program to accomplish for you, analyzing the characteristics of the programs available, and deciding which one will meet your needs.

SUSTAINABLE WINEGROWING AND MARKETING WINE

Up until now, the sustainable winegrowing programs of which I am aware have been dedicated to education and self-improvement, with regions and states coming together to develop programs to help themselves move along the sustainable farming continuum. However, there is an increasing interest in marketing wine based on the sustainable practices used to grow the grapes.

While marketing is essential to a sustainable business, the process can have a very different feel to it than the process of self-improvement in vineyard and winery management. The goal of sustainable winegrowing is to leave a smaller environmental footprint while contributing in a positive way to the local community. The goal of marketing is to add value to and sell more winegrapes and wine. While there is no code of ethics for marketing, in my opinion the wine industry should market sustainability based on facts, and resist the temptation to overstate the virtues of one green strategy or another. Moreover, the fact that sustainable winegrowing is such an all-encompassing paradigm provides some significant challenges to developing a simple, yet meaningful marketing message. Many fear that greenwashing is a real threat.

What is greenwashing? Wikipedia defines it as "a term that is used to describe the act of misleading consumers regarding the environmental practices of a company or the environmental benefits of a product or service" *(www.wikipedia.com)*. Evidently greenwashing was coined by suburban New York environmentalist Jay Westerveld in 1986, in an essay regarding the hotel industry's practice of placing green placards in each room, promoting reuse of guest-towels, ostensibly to "save the environment." He noted that in most cases little or no effort toward waste recycling was being implemented by these hotels and concluded that the actual objective of this "green campaign" on the part of many hoteliers was, in fact, profit increase. Hence this and other outwardly environmentally-conscientious acts that have a greater, underlying purpose of profit increase was "greenwashing" in his estimation (www.wikipedia.com).

MAKING A COMPLICATED TOPIC SIMPLE

It is clear that the consumer is paying more attention to a sustainable marketing message even though many are confused as to what the term "sustainable" means. As mentioned above, the temptation for marketing might be to focus on one aspect of sustainability instead of the complete picture.

Packaging is one example. There is wine in a box/Tetra paks vs. glass bottles vs. PET bottles, synthetic closures vs. corks, recyclable packaging, and so forth (Figure 15-3). From an energy standpoint, the argument for producing wine in a box, or wine in plastic bottles is compelling. Glass is heavy and one has to burn a lot of carbon-based fuel to move it around, so if wine is put in a box or a PET bottle, one is leaving a much smaller carbon footprint regarding transportation. Furthermore, producing glass creates about 45% of the carbon dioxide produced in the winemaking process, which includes transporting the grapes and the bottles to the winery.

While very important and useful, the carbon footprint concept is very complex and should be dealt with as such. It is an extremely popular topic at the moment, given the attention that climate change has attracted and, on the surface, appears to be something that can be discussed in simple terms. However, there is more to it than meets the eye. For example, it has been asserted by some that the greatest climate impact from the wine supply chain comes from transportation. However, transportation accounts for less than 20% of the CO^2 production in the winemaking process vs. the 45% from glass manufacturing mentioned above. Moreover, greenhouse gas emissions per bottle of wine shipped depends on how it is shipped and for how many miles. The most efficient form of shipping, in relation to fuel

Figure 15-3 Wine in a Tetra pak.

consumption, is by boat and the least is by truck, differing by a factor of over 700, and rail is much more efficient than trucking as well. Is buying wine from New Zealand in a PET bottle more sustainable that buying wine in a glass bottle from someplace in the US? This is a complicated issue.

While the carbon footprint is a great yardstick and deceptively easy to talk about, this measure is only a portion of the complete sustainability picture. Greenhouse gas production does not give us a measure of a company's commitment to things such as landscape planning, human resources, water use efficiency, water quality, air quality, or wildlife habitat enhancement.

While on the topic of carbon footprints, if a company claims it is carbon neutral, the only way this can be possible is if it purchases carbon credits or has a solar array that produces as much energy as is consumed in the process of wine production, from growing the grapes to the final delivery of the bottle to the retailer. If anyone says they are carbon neutral in part because their vineyards are sequestering carbon, it turns out that not enough is known yet about the magnitude of carbon sequestration by vineyards to make this claim. So being carbon neutral may not be a good measure of the carbon footprint of a company if they are achieving it mainly through purchasing carbon credits. It is important to look at a company's carbon budget and to see how they achieved carbon neutrality, not just look at the bottom line.

Figure 15-4 Logo for the Lodi Rules for Sustainable Winegrowing Certification program.

GREEN CREDENTIALS VS. GREENWASHING

Sustainable winegrowing is a positive and proactive experience, and if we are going to use it in marketing we should all make an effort to do so in a manner that is a reflection of what it is.

Many wine regions around the US have devoted years and large amounts of dollars developing and implementing extremely successful sustainable winegrowing programs. Here are some examples. The sustainable winegrowing program developed by the Lodi Winegrape Commission began in 1992. They developed one of the leading grower outreach and education programs focusing on sustainable winegrowing; they published and implemented the *Lodi Winegrower's Workbook* in 2000, and launched the *Lodi Rules for Sustainable Winegrowing* (*Lodi Rules*) third-party certification program in 2003, and by 2010 more than 10% of their vineyard acres (15,000 acres) have achieved certification.[9] The Central Coast Vineyard Team (CCVT) published and implemented California's first sustainable winegrowing self-assessment workbook in 1996 and have also developed an outstanding grower education and outreach program focusing on sustainable winegrowing, including the Sustainable Ag Expo for the last 3 years.[10] In 2008 they launched Sustainability

---

9. *www.lodiwine.com and www.lodirules.com*

10. *www.vineyardteam.org*

in Practice (SIP), their own sustainable winegrowing certification program.[11] The Napa Valley Sustainable Winegrowers Group formed in 1995 and has held regular meetings on sustainable winegrowing and published an IPM guide with regional specific information.[12] The California Association of Winegrape Growers (CAWG) and the Wine Institute joined forces in 2001 to form the California Sustainable Winegrowing Alliance to implement the *Code of Sustainable Winegrowing Practices Workbook* program which has had huge penetration into the communities of California winegrape growers and wine makers.[13] One of their many accomplishments was to benchmark sustainable practices in a report in 2004 and then report on the progress against those benchmarks in 2010. Fish Friendly Farming is a program in the north coast area of California that has been going since 1999 and is a certification program for vineyard properties that are managed to restore fish and wildlife habitat and improve water quality.[14] Washington State created *VineWise*, their online self-assessment workbook, in 2003.[15] In 2008, after several years of careful development, New York State launched the *NY Guide to Sustainable Viticulture Practices*.[16] And finally, the outstanding sustainable winegrowing program Oregon LIVE, certified by the International Organization of Biological Control, has been going at least since 1999.[17]

CONCLUSION

Sustainability is complex and does not lend itself well to sound bites. However, if the term sustainability is to survive as a meaningful label in the marketplace we must try to craft a simple message that is an accurate representation of what it is. Otherwise, the word will simply become one that companies can use to mean whatever they choose. If the term sustainability is to survive as a meaningful concept in the vineyard, then we need to meet the three challenges of implementing sustainable winegrowing described in Chapter 1; growing winegrapes in a way that is environ-

---

11. *www.sipthegoodlife.org*
12. *www.nswg.org*
13. *www.sustainablewinegrowing.org*
14. *www.fishfriendlyfarming.org*
15. *www.vinewise.org*
16. *www.vinebalance.com/*
17. *www.liveinc.org*

mental sound, economically viable and socially equitable. I hope the information in this book will help the reader meet those challenges.

One of the simplest, yet most powerful descriptions of sustainable winegrowing is the three E's or the triple bottom line. It is growing winegrapes and making wine that is environmentally sound, socially equitable and economically viable. This triple bottom line is also what makes sustainability unique among other farming paradigms. We should not expect any grower to have the "perfect" sustainable program, because it is simply not possible to grow grapes and make wine to sell without leaving an environmental footprint. I think we all have to come to grips with this fact and realize the goal is continual improvement. The goal is to move as far along the sustainability continuum as we are able to and tell our customers the story of how we are doing it. It is keeping our eyes on the receding horizon in the world of sustainable farming.

Figure 15-5   Vineyards near the foothills of the Sierra Nevada near Clements, California.

Resources

BOOKS

Keller, M. 2010. The *Science of Grapevines: Anatomy and Physiology.* Academic Press, San Diego. 369pp.

Smart, R., and M. Robinson. 1991. *Sunlight into Wine: A Handbook for Winegrape Canopy Management.* Winetitles, Adelaide. 88pp.

Skelton, S. 2009. *Viticulture—An Introduction to Commercial Grape Growing and Wine Production.* Published by the author

Wagner, P. M. 1996. A *Wine-Grower's Guide.* The Wine Appreciation Guild, San Francisco. 240pp.

Winkler, A. J., J. A. Cook, W. M. Kliewer, and L. A. Lider. 1974. *General Viticulture Second* revised edition. University of California Press, Berkeley

Wolf, T. et. al. 2008. *Wine Grape Production Guide for Eastern North America.* Natural Resource, Agriculture, and Engineering Service. 336 pp.

WEB RESOURCES

University of California Davis Integrated Viticulture Online:
ucanr.org/sites/intvit/

National Online Viticulture Resource:
www.eviticulture.org

National Cooperative Extension eXtension for grapes:
www.extension.org/grapes

Index

Handbooks for the Vineyard and Winery from The Wine Appreciation Guild

Concepts in Wine Technology
Yair Margalit

This detailed how-to guide, by physical chemist and winemaker Yair Margalit, is organized in the sequence of winemaking: in the vineyard: proper maturation, soil and climate, bunch health, vineyard disease states and grape varieties; pre-harvest: vineyard management and preparing the winery for harvest; harvesting: destemming, crushing and skin contact as it applies to both red and white grapes to pressing, must correction and temperature control; and finally fermentation and cellar operations.

Retail $39.95. ISBN: 978-1-934259-46-7
Paperback, 7 x 10 inches, 263pp, illustrated, charts, graphs,
    index.

Winery Technology & Operations:
A handbook for small wineries
Yair Margalit

The predecessor to Dr. Margalit's brilliant Concepts in Wine Technology is still one of the most widely used academic text books on the technology of winemaking in print. With exquisite detail, the book outlines every level of the vinification process, from the vineyard to the cellar.

Retail $29.95. ISBN: 0-932664-66-0
Paperback, 6x9 inches, 224pp, illustrated, index, appendixes

Concepts in Wine Chemistry
Yair Margalit

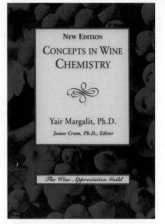

The primary text since for scores of universities and winemakers in a dozen countries, Concepts in Wine Chemistry, by physical chemist and winemaker Yair Margalit, details the basic and advanced chemistry behind the practical concepts of winemaking: must and wine composition, fermentation, phenolic compounds, aroma and flavor, oxidation and wine aging, oak products, sulfur dioxide, cellar processes and wine faults. Dr. Margalit also gives the biochemist's slant on the contentious question: is wine good for you?

$89.95. ISBN: 978-1-934259-48-1
Paperback, 7 x 10 inches, 446pp., illustrated, charts and graphs, fully indexed.

A Wine-Growers Guide
Philip M. Wagner

The best-selling practical "how to grow wine grapes" guide in print, this classic text has been the mainstay of both profession and dilettante vineyardists for more than a decade. The book includes information on propagating, planting, training and pruning, vines; and vine ailments and vineyard management.

$19.95. ISBN: 978-0-932664-92-1
Paperback, 5 ½ x 8 ½ inches, 240pp, illustrated.

Understanding Wine Technology, 3rd Edition
David Bird, Foreword by Hugh Johnson

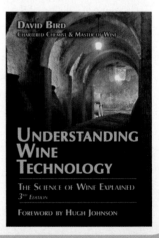

Any student who's ever logged credits in a viticulture and enology class knows Bird's book. It's the most widely assigned wine-science primer in the English speaking world. This completely revised and updated edition to Bird's classic textbook deciphers all the new scientific advances that have cropped up in the last several years, and conveys them in his typically clear and plainspoken style that renders even the densest subject matter freshman-friendly.

$44.95. ISBN: 978-1-934259-60-3
Paperback, 6 x 8 inches, 328pp., full-color illustrations and charts, fully indexed.

Wine Faults: Causes, Effects, Cures
John Hudelson, Ph.D.

A precise and comprehensive description of the problems encountered at times by all winemakers and wine judges. Every microbial infection found in today's wineries is fully described and arrayed in full color slides. Dense as the material may seem, the book is written in a manner that the layperson, or even the quality control professional who forgot that he ever took Organic Chemistry, can understand.

$39.95. ISBN: 978-1-934259-63-4
Paperback, 81/2 x 11 inches, 96 pp., full-color illustrations,
    fully indexed.

Biodynamic Wine, Demystified
Nicholas Joly, Foreword by Mike Benziger

Joly shares the core philosophy behind biodynamic viticulture and why such practices result in wines of regional distinction. He explains why the use of foreign substances like pesticides and fertilizers in the vineyard, and aromatic yeasts and enzymes in the cellar, disrupt vineyard ecology and are ultimately counterproductive to a wine's best, consistent expression.

$24.95. ISBN13: 978-1-934259-02-3
Paperback, 6x9 inches, color plates, fully indexed

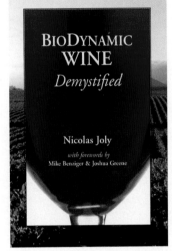